Biomedical Signals and Systems

Synthesis Lectures on Biomedical Engineering

Editor
John D. Enderle, *University of Connecticut*

Lectures in Biomedical Engineering will be comprised of 75- to 150-page publications on advanced and state-of-the-art topics that span the field of biomedical engineering, from the atom and molecule to large diagnostic equipment. Each lecture covers, for that topic, the fundamental principles in a unified manner, develops underlying concepts needed for sequential material, and progresses to more advanced topics. Computer software and multimedia, when appropriate and available, are included for simulation, computation, visualization and design. The authors selected to write the lectures are leading experts on the subject who have extensive background in theory, application and design. The series is designed to meet the demands of the 21st century technology and the rapid advancements in the all-encompassing field of biomedical engineering that includes biochemical processes, biomaterials, biomechanics, bioinstrumentation, physiological modeling, biosignal processing, bioinformatics, biocomplexity, medical and molecular imaging, rehabilitation engineering, biomimetic nano-electrokinetics, biosensors, biotechnology, clinical engineering, biomedical devices, drug discovery and delivery systems, tissue engineering, proteomics, functional genomics, and molecular and cellular engineering.

Biomedical Signals and Systems
Joseph V. Tranquillo
2013

Health Care Engineering, Part I: Clinical Engineering and Technology Management
Monique Frize
2013

Health Care Engineering, Part II: Research and Development in the Health Care Environment
Monique Frize
2013

Computer-aided Detection of Architectural Distortion in Prior Mammograms of Interval Cancer
Shantanu Banik, Rangaraj M. Rangayyan, and J.E. Leo Desautels
2013

iv

Models of Horizontal Eye Movements, Part II: A 3rd Order Linear Saccade Model
John D. Enderle and Wei Zhou
2010

Models of Horizontal Eye Movements, Part I: Early Models of Saccades and Smooth Pursuit
John D. Enderle
2010

The Graph Theoretical Approach in Brain Functional Networks: Theory and Applications
Fabrizio De Vico Fallani and Fabio Babiloni
2010

Biomedical Technology Assessment: The 3Q Method
Phillip Weinfurt
2010

Strategic Health Technology Incorporation
Binseng Wang
2009

Phonocardiography Signal Processing
Abbas K. Abbas and Rasha Bassam
2009

Introduction to Biomedical Engineering: Biomechanics and Bioelectricity - Part II
Douglas A. Christensen
2009

Introduction to Biomedical Engineering: Biomechanics and Bioelectricity - Part I
Douglas A. Christensen
2009

Landmarking and Segmentation of 3D CT Images
Shantanu Banik, Rangaraj M. Rangayyan, and Graham S. Boag
2009

Basic Feedback Controls in Biomedicine
Charles S. Lessard
2009

Understanding Atrial Fibrillation: The Signal Processing Contribution, Part I
Luca Mainardi, Leif Sörnmo, and Sergio Cerutti
2008

Understanding Atrial Fibrillation: The Signal Processing Contribution, Part II
Luca Mainardi, Leif Sörnmo, and Sergio Cerutti
2008

Biomedical Signals and Systems

Joseph V. Tranquillo

ISBN: 978-3-031-00531-2   paperback
ISBN: 978-3-031-01659-2   ebook

DOI 10.1007/978-3-031-01659-2

A Publication in the Springer series
SYNTHESIS LECTURES ON BIOMEDICAL ENGINEERING

Lecture #52
Series Editor: John D. Enderle, *University of Connecticut*
Series ISSN
Synthesis Lectures on Biomedical Engineering
Print 1930-0328   Electronic 1930-0336

Biomedical Signals and Systems

Joseph V. Tranquillo
Bucknell University

SYNTHESIS LECTURES ON BIOMEDICAL ENGINEERING #52

ABSTRACT

Biomedical Signals and Systems is meant to accompany a one-semester undergraduate signals and systems course. It may also serve as a quick-start for graduate students or faculty interested in how signals and systems techniques can be applied to living systems. The biological nature of the examples allows for systems thinking to be applied to electrical, mechanical, fluid, chemical, thermal and even optical systems. Each chapter focuses on a topic from classic signals and systems theory: System block diagrams, mathematical models, transforms, stability, feedback, system response, control, time and frequency analysis and filters. Embedded within each chapter are examples from the biological world, ranging from medical devices to cell and molecular biology. While the focus of the book is on the theory of analog signals and systems, many chapters also introduce the corresponding topics in the digital realm. Although some derivations appear, the focus is on the concepts and how to apply them. Throughout the text, systems vocabulary is introduced which will allow the reader to read more advanced literature and communicate with scientist and engineers. Homework and Matlab simulation exercises are presented at the end of each chapter and challenge readers to not only perform calculations and simulations but also to recognize the real-world signals and systems around them.

KEYWORDS

biological systems, biological signals, system response, stability, feedback, medical devices, PID control, filters, Laplace transform, correlations, convolution, signal processing, filters

Contents

Preface

When I learned signals and systems it was from an electrical engineer. I did well enough in the class because I learned to think in terms of electrical systems. But when I ran into mechanical or chemical systems I was lost. I knew that somehow they were related but I had gotten stuck so far down the electrical track that I couldn't find my way out. My goal is for you to be able to apply the concepts of signals and systems to *any* system. I will therefore do my best to introduce topics from multiple points of view. In that regard, biological and biomedical systems are very well-suited to the task—naturally provide a way to jump between mechanical, electrical, thermal, chemical, even optical viewpoints, and to span the range from medical devices to cell and molecular dynamics. The goal of this text is therefore to give you a self-guided tour through the major topics in signals and systems in a way that will allow you to apply concepts to your own area of interest.

WHY SHOULD I LEARN ABOUT SIGNALS AND SYSTEMS?

There are many sub-disciplines of engineering and so one may ask what it is that binds them all together, and there are a number of possible answers. I offer my own personal definition: An engineer is someone who can design and build something, given specifications and constraints, that never existed before. That "something" is nearly always a system, whether it is a bridge, a computer algorithm, a chemical process, a widget, a service or a circuit. It should not come as a surprise then that the most common course in an engineering curriculum, regardless of sub-discipline, is a course in signals and systems. It may go under different names like "Dynamics" or "Process Control," but at the core of these courses are the same concepts. And it is those concepts and thought processes that you will find here. Signals and systems are a collection of tools, a way of thinking and a lens through which engineers view the world.

Because of the common theme of "systems thinking" in engineering, you can often listen to the terms someone uses and pick out if they are an engineer or not. Each engineering discipline is infused with terms that derive from signals and systems—Memory, Time Constants, Correlations, Filters, Feedback, Control, Gain, are a few examples. But we can also turn this around. How will someone know that you are a credible engineer? One way to convince them, say at an interview, is if you are using the right words in the right context. If you are not an engineer, learning signals and systems will help you understand how engineers think and how best to communicate with them. Notice that this is different than being able to walk-the-walk. For that you need to be able to *do* much more than just know what the terms mean—you need to be able to use them flexibly in many contexts. While this book can help you talk-the-talk, it is lab that will teach you to walk-the-walk.

HOW SHOULD I USE THE TEXT?

Biomedical Signals and Systems is not meant to replace in-class activities, projects, supplementary lectures or office hours. The purpose is to provide three things. First, to allow you to frame questions in the language of engineering. Second, to flip the classroom environment by putting some the most basic information, as well as some of the technical details, in a text so that class time is freed up for more interactive and advanced material. Third, and maybe most important, is to highlight the connections between concepts that often get lost in a lecture.

There is yet one more way to use the text. I kept many of my old college texts and notes, whether I liked the class or not. Although this first exposure may not have always been ideal, the notes and text always feel more comfortable because I studied them inside and out at some point in my career. I find that when I look at them, somehow the material all comes rushing back to me in a way that a new text could never do. I hope that for some of you, these notes may serve that purpose someday.

HOW IS THIS BOOK DIFFERENT?

As you have probably already noticed, I am not trying to impress you with my writing style. There will be sections that will seem much more casual and less technical than most other textbooks. This informality allows me to include analogies and stories that will help you understand the material at a deeper level. My goal is not for you to become a theory expert, but rather to know where, when, and how to apply the theory. You can think of this book as an extended commercial for the exciting ways signals and systems can be used.

This decision comes at a cost. By removing some of the hard-hitting theory, you will not be able to participate in cutting-edge research in signals and systems. To dig deeper you may want to consult one of the excellent texts below:

Signals and Systems by Oppenheim and Willsky
Signals and Systems by Oppenheim
Schaum's Outline of Signals and Systems by Hsu
Signals and Systems using Matlab by Chaparro
Modern Control Systems by Dorf and Bishop
Signals and Systems Made Ridiculously Simple by Karu
Signals and Systems for Bioengineers by Semmlow
Signals and Systems in Biomedical Engineering by Devasahayam

LEARNING STYLES

There has been a great amount of research that has shown that we don't all learn the same way. After all, you aren't a robot but an individual (right?). Some of these learning styles are verbal, reading, writing, graphical and kinesthetic but there are probably more. You probably learn best from one or two of these styles, but it is often learning in a style that is hard for you that gives you a deeper insight. Perhaps you already have perfected learning by reading a text, in which case you should feel right at home. Personally, I learn best by writing down ideas to form connections and by being active. But I have learned some topics much more deeply by reading about them. So, if you too learn best in some other mode, try to bend the lessons in this text toward your learning style.

WHAT'S TO COME?

As a quick preview I want to explain how this text is organized. In most sections there is an introduction that gives some non-mathematical overview. We then dive into the theory, sometimes deriving important formulas. You could, however, read this text and still get something out of it even if you ignored all of the math. But to understand the engineering perspective, you will need to dig into the mathematical nature of signals and systems. To help you navigate the math, differential equations will be reviewed and a number of supplementary topics are addressed in appendices. In addition, the Matlab simulation environment will be used in some examples, but template script code will be provided. Embedded throughout are examples that highlight the concepts being discussed.

At the end of each chapter are some problems. These problems can be thought of as having dimensions to them. One-dimensional problems will require you to *use* a particular topic that was introduced in the sections. For example, we may have introduced the idea of a time constant and there will be some questions specifically on finding a time constant from a plot. Two-dimensional problems will require you to *integrate* information from the current chapter to a concept that was previously introduced. For example, you may need to relate a time constant to the stability of a system. Three-dimensional problems will require you to consider how the signals and systems concepts might be *applied* to other courses in your curriculum. You will also be challenged to think about how signals and systems might be applied to non-technical fields. You might be surprised to know that some anthropologists, sociologists, entrepreneurs, historians, bankers and lawyers use the same concepts in their own fields.

Joseph V. Tranquillo
December 2013

Acknowledgments

I first must thank the many students who have used early drafts of this text in Bucknell's Biomedical Signals and Systems course. I also must thank Jim Baish and Marie Jacob for all of the improvements they have suggested. Lastly I wish to thank my two children Laura and Paul, as well as my wife, Lisa, for encouraging me to finish this text.

Joseph V. Tranquillo
December 2013

CHAPTER 1

Introduction

1.1 WHAT IS A SYSTEM?

A rich definition of what a system is will unfold as you read this text. But, for the sake of getting started, we will adopt a working definition that is broad and then whittle it down to the engineering definition as we go.

A system is any collection of stuff that has been grouped together in some fashion.

If you take this definition at face value, you will realize that just about everything can be considered a system. The dewy decimal *system*, the planetary *system*, the democratic political *system* and a stereo *system* all fit our definition. These are obvious examples because they have "system" right in their name. And notice that some of these systems are groups of physical objects (e.g., planets) while others are groups of ideas (e.g., political systems).

    A more subtle type of system is one that describes an orderly sequence of events or a process (e.g., recipe, algorithm). A food recipe is simply a collection of directions for transforming raw ingredients into something delicious. You may also see systems disguised using the word *method*. For example, the Pilates *method*, *method* acting, and the LU decomposition *method* are either collections of ideas, movements or processes.

    In this text we will focus on physical systems, but many of the concepts can be applied to other fields. For example, the concepts of system stability, time constants, memory and frequencies that you will learn about later work perfectly well in the fields of economics, politics or psychology.

    One very important thing to realize about any system is that it is an abstract thing created by a human. This is not to say that the objects that make up the planetary system were created by humans, but the grouping was. Someone decided that a certain group of planets that are close to our sun seem to go together. Whether they were right or wrong can be debated (e.g., astronomers debating for a long time about whether Pluto is a planet or not), but most systems are collections that were created out of convenience—we can say *the solar system*, rather than list all of the planets. But the key here is that *you* get to decide what goes into the system. In fact, it is often by viewing a system in a different way that breakthroughs occur. For example, Sigmund Freud made many breakthroughs in psychology because he viewed the human mind as a system. By sending something into the system (a question) and observing what comes out (an answer from the patient), a psychoanalyst can determine something about what is happening in the system (the mind).

1.1.1   CAUSE AND EFFECT

We often talk about some event being *caused* by some other event. In our statements there is a clear idea of what came first in time—the cause always precedes the effect. For example, everyone agrees that a failed O-ring seal was the cause of the Challenger disaster. This failure existed *before* the disaster. If things happened the other way around (i.e., the explosion of the Challenger caused the O-ring to be burned up), we would switch our statement to say that the Challenger explosion *caused* the O-ring to burn up. The key idea here is that we assign the words "cause" and "effect" based on when an event happened in time.

What we are talking about here is the idea of a *causality*. The *belief* in causality is one of the two central pillars upon which all of science is built. The other pillar is that the rules of *logic* are self-consistent and if applied properly will yield truth. Although some highly technical mathematical derivations demonstrate that we cannot really talk about how to prove if something is true or not (Godel's incompleteness theorem), the fact that we are able to manipulate our environment and create self-consistent theories is a testament to the fact that these two pillars are sturdy in practice.

In a system that describes a process, the idea of cause and effect is natural because the system is by definition a sequence of events that follow each other one after another in time. For a grouping of ideas or objects, the cause and effect can become much less obvious. In some systems, like the planets, there is no clear cause or effect. In other words, Venus did not *cause* Mars to exist. They are on equal footing if our system is defined as only the planets. The collection of planets is *acausal* because there are no causes or effects. But when we turn to dynamic systems the possibility for cause and effect arises. The planets are dynamically moving, and they do have an effect on one another through mutual gravitational influences. But it is still a bit strange to say that their motions cause one another. Here some would call this a *partially causal* system—there is influence but no direct causes. Many complex systems, such as a biological system, are partially causal in that cause and effect become blurry. The parts of the system are so interdependent that it is hard to tell what is causing what.

In a true causal system the timing helps us differentiate between causes and effects. For these systems we will introduce the terms *input* and *output*. When you hear or see input, think cause. When you hear or see output, think effect. These will be the types of systems we will consider.

We need to be a bit careful here. Think about what would happen if I threw an eraser at you (which I would never do, of course). You would react in some way to that eraser after I started the motion to throw it. So the throwing of the eraser would be an input because it caused a reaction—anticipation. There are times that a system can use information to *anticipate* a reaction. We know that biological systems do this. For example, if you are playing second base, you *ready yourself* for the possibility of a ball coming your way. But you only do this based on cues, such as the pitcher winding up or a batter taking a certain stance in the box. Non-biological systems can anticipate (or *forecast*) too, but we do not consider this to be non-causal behavior. In this text we are going to consider only causal systems, but you will find that some of our systems (especially in the chapter on control) will in fact have built-in mechanisms for anticipation.

There is still a fourth possibility and that is *non-causal* systems. There are three possibilities. The first is that two events occur but are not connected by cause and effect. Instead they are related by their meaning. Often this is interpreted as there being a logical cause and effect without there needing to be time taken into account. This is the idea behind Carl Jung's Synchronicity. Jung used Synchronicity to explain, among other things, a number of pseudo-scientific phenomenon (i.e., ESP, strange coincidences). The second possibility is that an effect can precede the cause. Modern science tends to associate these types of systems with the occult. For example, reading tea leaves, astrology, lines on hands (phrenology) and other fortune-telling methods claim to be able to see effects before the causes have occurred. The third example is what is called a *degenerate* or *null* system. In this type of system there can be an effect without a cause. When we say something like "this occurred for no reason," and truly believe it, we are really saying that there is no cause for some event. It is important to point out that since the science upon which engineering is built is itself built upon the idea of causal systems, it by definition will dismiss (or possibly not even consider) any system which is non-causal. In fact, it could be argued that as soon as some system can be described in the language of cause and effect (leading to prediction), it has entered the realm of science.

1.1.2 THE SYSTEMS OF ENGINEERING

We have established that a system is a word used in many fields to describe collections of physical objects, ideas or processes. But what about engineers? What does an engineer mean when they say *system*? In short, we nearly always mean either a physical system or a plan (design) for a physical system. In other words, engineers generally do not design political systems or discover scientific theories; we deal with the physical world. We do however sometimes design processes. For example, an engineer might be charged with the task of reorganizing how parts for a biomedical device flow through a factory. The individual machines may be exactly the same, but the flow of the product may change. The field of study that is concerned with these types of problems is called industrial engineering (or sometimes operations research). For this text, however, we are going to focus on designing a physical system.

1.2 WHAT IS A SIGNAL?

In our systems definition, we used a very important phrase: "grouped together in some way." In some systems, this grouping may be very loose. But in most coherent systems, there is something that clearly connects one part of the system to other parts of the system. Again, the connection may be an idea or something more physical, but for an engineer this connection is almost always something physical (e.g., a wire that carries current, a tube that carries flow). The connection enables the flow of information (e.g., voltage in a wire, pressure in a tube). And that information may be encoded in many different ways: mechanical, electrical, chemical, thermal or optical. The key point here is that the different parts of the system interact by exchanging *information* using some physical quantity in the system.

A signal is information that is exchanged between different parts of a system.
Systems use signals to communicate.

So why not skip the signals and just get to the systems part? As an engineer you will find that you cannot measure a system! What you can measure are signals generated by a system. In fact, one way to differentiate a system from a signal is that a signal can be measured (e.g., color, temperature, voltage). As such, signals contain information about the properties of a system.

There are two different ways to think about a signal. The first is as a property of a system. In other words, you can't measure a planet, but you can measure the properties (e.g., diameter, distance from sun, color, temperature) of a planet. Although some might consider these signals, engineers usually do not. A system *property* that does not change, is often called a *constant* because no information is being exchanged. When engineers use the word signal, they almost always mean some measurable property of a system that *changes* over time. It is a dynamic property of a system and is information rich.

We can broadly classify engineering signals into three types. First are *inputs* to a system. Input signals are special because they originate outside of the system boundary. Here, the caffeine in your coffee or tea would be a good example of an input to your system from the outside. Second are *outputs* from a system. The outputs are measured outside the system boundary. Caffeine is broken down inside the body and then leaves the body. So how do we tell the difference between an input and an output? Here is where the idea of a causality is important. An input can be thought of as a cause that *acts on* a system. An output is the *effect on* or *product of* the system. Using our idea of causality, it is therefore easy to distinguish an input from an output because the input must always precede the output. Third, there are *internal* signals by which different parts of the system send information internally. For example, in your body your endocrine system may send hormones to your heart through the blood. Here you can measure the hormone levels, so that is the signal, but that signal remains entirely contained within the system boundary (you). Despite the fact that this signal lies entirely within the system boundary, engineers have developed sensors that can read these internal signals.

1.2.1   SIGNALS IN ENGINEERING

When you encounter a plot that has the generic axes in Figure 1.1, you are almost certainly looking at a signal (internal, input or output). You should note that it is possible to have a signal where the bottom axis is space (like a distance) or some other variable. Instead of a signal, an engineer would often call this a *distribution* or in two dimensions, an *image*. We will not consider these types of signals, but all of the techniques we will build up will apply to these other types of information.

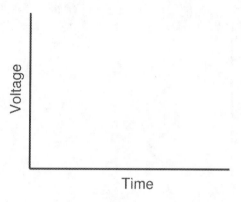

Figure 1.1: Signals in engineering.

1.2.2 SENSORS

In many modern engineering applications, the information in a signal, whether it is chemical, mechanical, fluid or thermal, is often represented electrically as a voltage. The simple reason is that circuits and computers are powerful tools that are used in almost every field. Since the medium of both circuits and computers is electricity, it has become natural for engineers to translate important signals they wish to measure into voltages. For example, if an engineer wishes to record the pH of blood over time, a chemical reaction is used to translate changes in pH to changes in voltage. This chemical-to-electrical translation is typically performed by a device called a *transducer*. The traditional definition of a transducer is a device that transforms one type of *energy* into another (e.g., chemical energy into electrical energy). For an engineer this definition is interesting but not useful. What is useful is that a transducer can convert a *signal* (e.g., mechanical, chemical) to an electrical signal that retains the information.

1.3 SYSTEM BOUNDARIES

Whenever the word "system" is used the world has been divided into two parts: 1) the system and 2) everything else. It is important to realize that the dividing line between 1 and 2 is not a physical thing and is only a definition that is made by a human. Where this division is made does not matter but clearly defining how the division was made can be important. To demonstrate the point we will use the idea of a *block diagram* (which will be formally introduced in Chapter 5).

In Figure 1.3, we see a graphical representation of a generic system. Systems are represented with blocks (usually a rectangle). Signals are represented with arrows. Input signals are arrows that point into the system, while output signals point out of the system. The convention of blocks and arrows is used throughout engineering.

In Figure 1.3 we have one large system (SYS1) that is composed of two smaller systems (SYS2 and SYS3). If we consider the signal *a*, it is external to both SYS1 and SYS2 and therefore

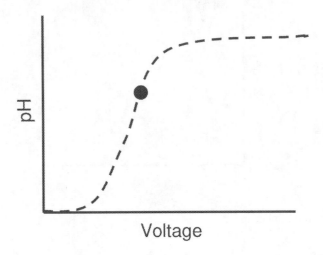

Figure 1.2: Generic calibration curve.

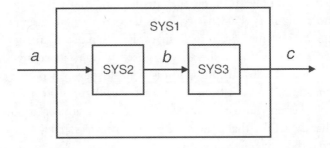

Figure 1.3: System boundaries.

an input (note the direction of the arrow) to both. If, however, we consider signal b, we have a dilemma. From the perspective of SYS1 it is internal, from SYS2 it is an output and from SYS3 it is an input. So which is it? In Section 1.1 the inputs and outputs were determined by when they occurred. But here b (the input signal) occurs at the same time as b (the output signal) and at the same time as b (the internal signal). If this sounds confusing that's because it is! The idea of causality isn't going to help us here. This dilemma occurred because the signal b, or any signal for that matter, is not an input, output or internal until we clearly define what the system is. In our example, once we define which system we want to analyze, the signal b takes on meaning. But a signal not being an input, output or internal on its own is important because it can act as all three at once (as we can see from this example). A little thought will show that it is this property that allows for groups in a system to be connected together. You will encounter many more examples where a single signal acts in many ways depending upon how you define your system. The bottom line is to define your system first and then consider the signals.

1.4   DESIGN USING SIGNALS AND SYSTEMS

The reason for engineers to take a course in signals and systems is because in the design of nearly any device, an engineer will encounter 1) What you have to work with, and 2) What you want to do with it. These sound very much like inputs and outputs! In fact, you can think of them in just this way. Design is a process for turning what we have into what we want. Typically what we want (output of our system) are called *products*. The inputs may be problems, time, money, resources, materials, current technology and people. For this reason, systems thinking is very much integral to design and there is a lot of cross-talk between the two ways of thinking.

CHAPTER 2

System Types

2.1 INTRODUCTION

The general view of systems from the last chapter is a good place to start, but by making our definition so inclusive it is hard to move forward. The usual way to proceed is to begin by making a few broad assumptions. For example, we will only work with causal systems because we gain some analytic power. In some instances this may be enough. In others, we will need to make some further assumptions. The purpose of this chapter is to narrow the focus to some specific types of systems that are analytically tractable.

2.2 CONSERVATIVE AND NON-CONSERVATIVE SYSTEMS

Engineers and scientists tend to work with what are assumed to be *conservative* systems. The system is not leaking or gaining anything that cannot be counted or measured. If we encounter a system that does seem to be leaking (or gaining) something, our first impulse is to go looking for it—we almost can't imagine a system that is truly non-conservative. This concept is at the heart of all conservation laws (e.g., charge, momentum, spin, matter/energy).

In this book we will make conservative systems our default. It is true that many systems dissipate energy, often as heat. From one viewpoint, we could ignore the loss of heat, in which case the system will seem to be energetically non-conservative. But, as soon as we account for the heat, it will be conservative again. The same can be said for systems where some other quantity other than energy is being considered. It all will depend upon how you draw the boundary around your system.

There are some other great examples of non-conservative systems. At some abstract level, your brain is not conservative—not every memory that goes into your brain can be taken back out—some memories are "leaky." Other examples are the ideas in books. The author may have written a finite number of words, but those words can then generate other ideas, other books, revolutions and so on. We will not concern ourselves with such systems, but they are good examples to ponder.

2.3 OPEN AND CLOSED SYSTEMS

The difference between *open* and *closed* systems is in the inputs and outputs. Closed systems have no inputs or outputs—they are isolated and operate entirely on their own. If a system is truly closed, it is a philosophical/mathematical/logical supposition that it would need to then be conservative. Making the assumption that a system is closed, then allows us to assume that everything is conserved.

In reality, all physical systems that we interact with are open—having inputs, outputs or both. We will often assume, however, that they only have a few inputs and outputs that matter. Of course *matter* is a fuzzy word, but it will take on meaning given the context.

2.4   STATIC AND DYNAMIC SYSTEMS

A *static* system does not change over time. For example, a rock may not change its weight (at least on the time scales we are used to) and the text of Shakespeare's *Romeo and Juliet* will (hopefully) be preserved over time. A *dynamic* system does change over time. Typically we analytically capture the dynamic nature of these systems in differential or difference equations, the topic of Chapter 3. It is important to recognize, however, that some disciplines will treat the same system as static, while for others it will be dynamic. A civil engineer might treat the location of a river as static, and thus build a bridge over it. A geologist, however, might treat the same river pathway as dynamic. In the same way the view of a system may depend on how the system boundary is drawn, whether a system is considered static or dynamic often depends on the time scales being considered.

2.5   CONTINUOUS AND DISCRETE SIGNALS AND SYSTEMS

Often in engineering, we make a big deal over the output of a dynamic system being continuous or discrete. To understand the difference, a dynamic system moves through various *states*. These states are snapshots of all the variables of the system at any given time. So the state of your blood is the current level of Calcium, Iron, Magnesium, Potassium and other ions, but also flow rates, white blood cell count, pH and many other variables. That is the state now and it may even be different in different places in your body. So to get a full snapshot we would need to know the quantities everywhere in your body. If we take a snapshot at some later time, some of those variables may have changed.

The number of states we need to fully characterize a system (i.e., making a complete snapshot) is the *order* of the system. The order of the system is based upon how many variables are needed to represent the system.

In a *continuous* system we assume that there are no jumps from one state to another. In a mathematical sense, if we think about what the state is now, and what it is in the next instant (where the next instant is infinitesimal Δt), we will get a smooth progression from one state to the next. Of course, we can represent this as a limit as Δt goes to 0 and arrive at differential equations. The rules for a continuous system are often modeled using differential equations. Often engineers will call a continuous system an *analog* system. Signals from a continuous system are called analog signals.

If there are higher-order derivatives, that is okay because we can always break them down into several first-order differential equations. So

$$\frac{d^3 y}{dt^3} \tag{2.1}$$

would be a third-order term, because it will take three first-order differential equations to represent Equation 2.1.

If we allow states to instantaneously jump from one value to another, without traversing the points in between, then the system is *discrete*. Engineers usually call discrete systems *digital* systems and model them using *difference* equations. Signals from a discrete system are called discrete signals.

A typical difference equation is a rule that shows how to get from past states to future states.

$$x[n] = x[n-1] + 3x[n-2] + x[n-4]^2 \tag{2.2}$$

A few things are worth noting. First, unlike continuous signals and systems, where variables are placed in parentheses, (), discrete systems use brakets, []. This is to signify that time jumps from one time to the next—from $n-1$ to n with no time in between. Second, to produce the current value, $x[n]$, the rule is to add together the last value of x to three times the value of x two time steps ago, and then add the square of the value of x four times ago. Once the calculations on the right hand side are performed, the value of $x[n]$ can be found. This process can be repeated over and over again to march the value of x forward in discrete steps.

This text will focus primarily on continuous systems but engineers deal with both analog and digital systems. Both have their advantages and disadvantages. For example, if a signal is digital, we can store and graphically represent it on a computer. But it can make certain operations occur more slowly than if they were done in an analog system (such as filtering a signal using an analog circuit). The concepts we will cover all have meaning for both analog and digital systems, and where appropriate we will explore the similarities and differences.

Because engineers work with both types of signals there are ways to convert between analog and digital signals. To gain some feel for how this conversion occurs we will discuss how an analog signal can be *sampled* and *quantized* to get to a digital signal. Engineers call the device that performs sampling an *analog to digital* (A/D) converter. The basic process is shown in Figure 2.1.

In sampling we will use a sensor and some hardware to only record the value of the signal at particular values of time. There is no requirement on how often to sample and it is possible to simply record as many values as fast as possible. Most often, however, we sample at a certain rate, known as the *sampling rate*. When a sampling rate is used, the time increments between readings stays the same—there is a *sampling period*. The values of the analog signals between these times are ignored. From the left hand side of Figure 2.1 you may intuitively realize that we need to

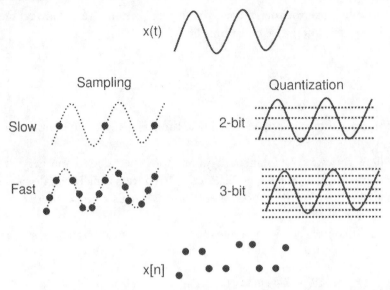

Figure 2.1: Analog to digital conversion using sampling and quantization.

sample at a rate fast enough to get a clear representation of the signal, a topic we will explore more in Chapter 11.

The right hand side of Figure 2.1 shows the other consideration in A/D converters and that is quantization. In the course of moving from one state to another, the signal shown might move through a continuous set of numbers. If the state moves from 3 to 4, at some point in time it will have passed through the value of π. But π is not a number that is easy to represent on a computer. In quantization, the hardware limits the values that can be recorded. If representing numbers between 3 and 4, the hardware may limit the *resolution* to 0.1. So 3.0, 3.1, 3.2 and so on can be resolved and any values between will simply be rounded to the nearest representable number.

The typical way engineers talk about quantization is to set the number of values allowed as powers of two. This is directly related to the fact that computers use base 2 (on-off) logic. Each power is a single *bit* of information. So, with one bit, only two levels can be determined. With two bits, four levels can be resolved (see Figure 2.1). And with three bits there are eight levels. In general the more bits used to represent values, the higher the resolution. The number of levels follows 2^b, where b is the number of bits. When an engineer says that an A/D converter is 8 bits, that means that the hardware can resolve 2^8 or 256 different values.

There is more to quantization than the number of values. You may have noticed that on the right hand side of Figure 2.1 the levels may or may not cover the full range of the analog signal. If the full range is not covered, any values higher (or lower) will be *clipped*—essentially rounded to the highest (or lowest) available value. Ideally the hardware will let you adjust these limits. So it is important to also specify the minimum and maximum values for an A/D converter. For example, an engineer might set an 8 bit A/D converter to record values over the range 0-5V. What this means is that the 0-5V range is cut up into 256 values. Because 0 and 5 must both be represented, in reality we have 255 divisions. To compute the resolution of the A/D converter

$$\frac{Signal\ Range}{2^d - 1} = \frac{5V}{2^8 - 1} = 0.0196V = 19.6mV \tag{2.3}$$

It is important to recognize that whenever sensors are used and the results are stored on a computer, at some level sampling and quantization are taking place. Often we are recording from more than one sensor at a time—perhaps to get a more complete picture of the system state. In these cases there are often dedicated A/D convertors for each signal, with each signal coming in on a *channel*. Recording from multiple channels at the same time is often called *Data Acquisition* (or DAQ for short). Some DAQ devices do more than just record sampled and quantized signals, and can also send out sampled and quantized signals to control devices.

It is also possible to move from a digital signal to an analog signal (D/A converter) by interpolating the continuous values that lie between the known discrete values. We will not consider D/A converters here.

2.6   STABLE AND UNSTABLE SYSTEMS

In dynamic systems, an important topic that we will consider is if the system is stable. This is a tricky topic to truly understand, so we will give a definition first and then add to it. A stable system is one where, as long as you don't try to break it (by sending in very large inputs), the output will remain bounded in some way (won't go to infinity). But there are many ways a system can stay within some bounds. The most obvious is that the system could settle into some state that does not change. This is a static equilibrium state. Without a *perturbation* (an input from the outside) the system will stay at that equilibrium.

There are a few others ways in which a system may be stable but where it *does* continue to change. For example, a system can have what is known as a *dynamic equilibrium*—it could be some repeating cycle (a ticking clock, known as a limit cycle) or some non-repeating, but never settling cycle (the chaotic weather). In both cases, the system is stable because it stays within some bounds. But at the same time it never settles down to one particular state.

We will discuss stability in much more detail in Chapter 6, but one point should be made here. When we find a system in nature, we almost always start by assuming that it is stable. If it were not, it wouldn't be around long enough for us to observe! Where the idea of stability is really important is when we, as engineers, are *designing* a system. We want to make sure that the system we design, and eventually build, will be stable. And in future chapters you will learn some analytical tools to predict the stability of an on-paper design before it is built.

2.7   TIME VARYING AND TIME INVARIANT SYSTEMS

Some systems vary in time and others do not. But we need to be careful here. This is not the same as static or dynamic—whether the *outputs* of the system change or not. A changing output gives an indication that the system is dynamic. The question of whether a system is time-varying or invariant is about whether the *system itself* is changing over time.

Now there may be lots of reasons why a system might change over time. It may age or become tired. Some relationships in the system might change due to adaptation. There is a practical implication here. For a time-invariant system, we can put in an input and get an output. If we wait some time and then put in the same input, we will get the same output. In a time-varying system the input-output pairs may change over time. There are many systems that we believe are time-invariant, for example the laws of physics (although there are some who think that some physical "constants" like the speed of light have changed over astronomical time scales). Again, we need to be a bit careful here. For all practical purposes, even if the laws of physics do change, they change so slowly that over our lifetime (and maybe even the history of our species) we can just assume the laws of physics to be time-invariant. You can, however, contrast this with the music on a CD. We would hope that the CD would play the exact same thing every time. But over time, a real CD might degrade over the course of your life and no longer produce the same output. On a short time scale things may appear to be time-invariant. For example, a hiking path is mostly invariant over the course of a few days. But on a long time scale it may be time varying—usage by many hikers, floods, tree falls and other events can change the path.

There are many social situations that are drastically time-varying. Think about investing in the stock market, or making a move to a new job at the wrong (or hopefully right) time. How about in the game of Monopoly where landing on Park Place *now* means you can buy it, but if you land on it ten moves from now you might need to pay rent. The Monopoly board is a time varying system.

In math speak, a time varying system looks like

$$a(t)\frac{dx}{dt} \tag{2.4}$$

Here the constant out front is $a(t)$ (not the constant a) because it is varying over time. If the system were

$$a\frac{dx}{dt} \tag{2.5}$$

then we would say the system was *time-invariant*. You may also hear engineers say that this system obeys the principle of *reciprocity*.

Because the time properties are inherent in the system, they are not tied to the inputs or outputs. So inputs to a system do not alter the time varying properties of a system. If $H(t)$ is the input to a system

$$a\frac{dx}{dt} = H(t) \tag{2.6}$$

it is still time-invariant.

Time varying systems can be difficult to handle so in this text we will always begin by assuming our system is time-invariant.

2.8   DETERMINISTIC AND NON-DETERMINISTIC SYSTEMS

Deterministic systems are those for which we could, at least in principle, write down rules that would predict what will happen next. We are assuming here that our system is dynamic. A truly deterministic system is the clock-work world imagined by Laplace (based upon his interpretation of Newton's Laws). In a nutshell, if we knew the values of all the variables and constants right now (the current system state), and all of the rules, we could figure out the entire history (past and future) of the system. This is a pretty strong statement, and one that many people have had a hard time accepting (e.g., Do we really have free-will?). We aren't going to get bogged down in this philosophical debate. What you need to know is that deterministic systems are governed by precise rules and we will assume this as a default throughout the book.

We can contrast a deterministic system with a *non-deterministic system*. In a non-deterministic system there is some part that is unknowable (or unpredictable) for some reason. It doesn't really matter why it is unknowable, and it doesn't matter how much of the system is unknowable. Just a little bit of the unknown will make the system non-deterministic. The most common way that something would be unknowable is if there is something inside the system that is *random*. In fact, what random means in this context is unpredictable—we don't know what a random system will do next and therefore there is no way to write down precise rules. Another way to think about this is that given one input, there are many possible outputs. Some think of biological systems as being non-deterministic because they seem to have lots of sources of random noise. As we will see in examples throughout the book, that is not necessarily the case.

Even if we did encounter a non-deterministic system, we wouldn't necessarily need to simply throw up our hands and give up. We can still define rules that are not set in stone. They can be probabilistic rules, sometimes called *heuristics* or *rules of thumb*. For example, if we are playing Black Jack, we might make it a fuzzy rule to ask for another card if our total is below 16, but stay if it is above. This isn't always going to work, and if we are on a hot streak we might bend our heuristic a bit. As an engineer, if the probabilities in a rule start to approach 1 (or 100% certainty) then we often assume that there isn't enough randomness to matter and we treat the system as deterministic. It is the assumption of a deterministic system that will lead us down the path where we can predict outputs given inputs. And we will be able to learn about the system that generated the input-output pair.

2.9   FINITE AND INFINITE SYSTEMS

If a system has a countable number of inputs, output and rules that govern how the system transforms inputs to output, then the system is *finite*. Otherwise, it is considered to be infinite. Typically we don't worry about the number of inputs and outputs, but rather focus on the number of rules. In this way finite systems are deterministic. It is a premise of most all of science that there are a countable number of rules that govern everything—thus the hope that we could someday discover "the laws of the universe" and that there will be some countable (hopefully small) number of rules. As stated earlier, in this text we will only consider rules that can be expressed as differential or difference equations. But the same logic of a countable number of rules governs all finite systems, regardless of how the rules are expressed. The number of rules needed to characterize a system is often called the system *degree* or system *order*.

In an infinite system, the degree (or number of rules needed to characterize the system) is not countable. Mathematically, an infinite system can generate what a mathematician would call a random signal and is therefore non-deterministic and unpredictable.

Engineers usually do not draw the line in the same place as a mathematician. If the number of rules becomes too great we will simply assume that for all practical purposes the system is infinite. The system generates pseudo-random signals and is still in practice unpredictable. Another way to say this is that if we see something that is statistically random, then we assume it is the output of an infinite system. Often engineers would say that this system is a *random process*. We will use this idea in the next section.

There are many biological systems that might appear to be infinite to an engineer—there would be essentially an uncountable number of rules to completely describe the whole system. But engineers have some tricks that can be used. It is often the case that some rules only apply in special situations—they can be ignored most of the time. By making some assumptions, it is possible to treat a biological system as if it were finite, and at this point it is possible to model the system with equations—the topic of Chapter 3.

2.10   LINEAR AND NON-LINEAR SYSTEMS

If a system is deterministic, it may be *linear* or *non-linear*. The difference can be summed up in a few simple equations. But before these are presented, we will try to build up an intuitive feel. If you walked into your favorite burger joint, they might sell you a single paddy burger for $1.50. If you add an additional paddy what would you expect? Should it be $3.00? In reality, it probably costs more like $2.00. Here is a fundamental difference between linear and non-linear systems. If you think that doubling the number of paddies should double the price, then you are like a linear system. If you think that two paddies should cost less ($2.00), then you are like a non-linear system. Just to drive the point home, we could keep going. If we had 6 or 7, or n paddies, how much will the price go up? If we made a plot of number of paddies on the x-axis and price on the y-axis, we would see something striking for a linear system—the plot would be a straight line, thus the term linear system. If the plot is non-linear it will be something other than a straight line. In fact, one day in grad school a group of us over lunch computed the cost for an infinite paddy burger at Wendy's—it came out to around $6.00—clearly non-linear.

We will make a big deal about linear systems throughout the text, because they have certain properties that make them much easier to deal with (e.g., solve analytically). We can now begin to dissect those properties. First, there is the property of *additivity*.

$$x(a + b) = x(a) + x(b) \qquad (2.7)$$

Here we have a system x, and we send in some input a and get some output $x(a)$. Then we send in another input b and get an output $x(b)$. The equation above says that it doesn't matter if we send a and b in together (left side) or separately (right side), we get the same result. This is a powerful statement that we will take full advantage of later on.

The other property is multiplicative (sometimes called *homogeneity*).

$$x(\gamma a) = \gamma x(a) \qquad (2.8)$$

Here we have some constant γ. On the left side, we have just made our input larger by γ. On the right side we have first sent a throughout the system and then multiplied by γ—it makes no difference because they are equal in a linear system. In other words, the output scales exactly with scaling the input.

These two properties mean that linear systems are fairly easy to predict. Once we know one input-output pair, we can predict a whole family of scaled and added versions of the system. It is often said that in linear systems, the whole is exactly equal to the sum of the parts.

You may have guessed that a non-linear system is any system that doesn't follow additivity or homogeneity. You are right! What is important here is that nearly every real system you will encounter is non-linear. Why is this? Non-linear systems can have all sorts of breakpoints. Engineers often call these phase transitions or bifurcations. These are places where the system will abruptly change its behavior. For example the phase change that occurs in materials as you change the temperature and pressure gives rise to phase diagrams. There are abrupt lines in these diagrams where a small change in temperature or pressure can completely change the make-up of the material—say from liquid to solid. But this is an example where a small change makes a big difference. In some cases we have just the opposite, a large change makes little difference. This idea is at the heart of the biological idea of homeostasis—even large perturbations are resisted. We often think of non-linear systems as being hard to predict. They are more (or less) than the sum of their parts.

It is now time to practice identifying linear and non-linear systems. Because of additivity, we can think of a linear system as composed of a collection of *linear terms*. So we are going to take a look at a number of terms and discuss their linear properties. Given the term

$$\frac{d^2 x}{dt^2} \tag{2.9}$$

we wish to know if it is linear or not. To find out, imagine this term as the rule that governs how inputs determine outputs, as in Figure 2.2.

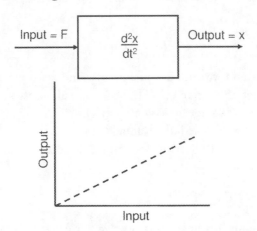

Figure 2.2: Assessing a linear system.

When any input is plotted against its output pair, the plot is a straight line and therefore a linear system. In fact, this is an empirical way to determine if a system is linear—send in many inputs and see if the outputs scale. What is more, the system in Figure 2.2 is in reality the well known formula $F = ma = m\frac{d^2 x}{dt^2}$, where a is the slope of the line and can be interpreted as an acceleration. Any higher-order derivative of a variable will be linear—the derivative is what is known as a *linear operator*. So

$$4\frac{d^3 y}{dt^3} \tag{2.10}$$

is linear because it is a higher-order derivative that has been multiplied by a constant. We can also consider

$$b(t)\frac{d^3 y}{dt^3} \tag{2.11}$$

which is time varying, but still linear. Now consider

$$x^2 \tag{2.12}$$

and imagine putting in an input of 3 with the output of 9. Now imagine doubling our input to 6. We get out 36! So doubling the input does not double the output and therefore this is a non-linear term. In fact, any powers (including square roots) will mess up linearity. So

$$\sqrt[3]{x} \tag{2.13}$$

$$\left(\frac{dz}{dt}\right)^3 \tag{2.14}$$

are both non-linear terms. Transcendental functions are also non-linear.

$$\sin(x) \tag{2.15}$$

$$\log(x) \tag{2.16}$$

The last way we can mess up linearity is if we have two different *independent variables*, say x and z, and we combine them together in some way other than addition or subtraction.

$$x^z \tag{2.17}$$

$$\frac{x}{z} \tag{2.18}$$

$$xz \tag{2.19}$$

are all non-linear. Now that the definition of a linear term is known we can assess some more complex systems.

$$y + \frac{dy}{dt} + 4\frac{d^2y}{dt^2} \tag{2.20}$$

$$u\frac{dy}{dt} \tag{2.21}$$

$$5\frac{d^2y}{dt^2} + \cos(y) \tag{2.22}$$

$$y + \frac{dy}{dt} + 4\frac{d^2y}{dt^2} - F(t) \tag{2.23}$$

$$\frac{dy}{dt} + 4y = \frac{d^2z}{dt^2} \tag{2.24}$$

The first equation is linear because we just have the sum of scaled higher-order derivatives. The second equation is tricky because we need more information. If u is a constant, then this is just a linear equation (property of homogeneity). But, if u is a variable of the system, then this equation is non-linear. The third equation is definitely non-linear because of the cosine term. The fourth equation is in fact linear. The term $F(t)$ (when written as a function that depends on time) is assumed to just be something from outside the system. It does not have any effect on linearity. The fifth equation is also linear, because all of the terms are linear. In other words, having terms that depend on another variable (like z) is okay, as long as that term is itself linear.

We will return to a slight modification to system four above.

$$y + \frac{dy}{dt} + 4\frac{d^2 y}{dt^2} - F(y) \tag{2.25}$$

Here the function F is dependent upon the variable y. What this means is that F is a subsystem of the larger system, similar to Figure 1.3. The F subsystem takes in the variable y and transforms it into $F(y)$. To know if the modified system above is linear or not, we need to know if F is a linear operator. If F obeys the properties of linearity, then the equation above will also be linear. If not, then it will cause the overall system to be non-linear.

As in many of the systems here, linearity is a strong assumption—all it takes is one term to make a system non-linear.

We have made a distinction between linear and non-linear systems as if the distinction were black or white. Analytically, that is true. But practically, it is more like a spectrum. For example, a resistor is usually thought of as a linear circuit element. The reason is because of Ohm's Law— $V = IR$ is linear because we can plot voltage versus the current and we will get a straight line. But that only works for a real resistor if you keep the current within a certain range. If the current is high enough, the resistor will melt and then no longer have the same linear properties. All physical systems become non-linear if pushed far enough.

Some systems are very strongly non-linear. Usually these systems have a lot of co-dependent variables. To anthropomorphize a bit, many variables "care" about what other variables are doing. But it is possible to have a weakly non-linear system where there is only one small mixing of independent variables or small non-linear function. Here we may be able to treat the system as if it were linear and then just allow a few exceptions for when it does not behave this way. In other words, there are some real systems that are close to linear.

Almost every physiological system, process or biomedical device is non-linear. So why should a biomedical engineer learn about linear systems? On the surface, being a credible engineer requires you to know the concepts and language of linear systems. But a deeper reason is that with a few tricks we can treat a non-linear system as if it were linear as long as we keep the inputs in a certain range. An engineer would call this a *linerarization* of a non-linear system. We will not discuss these techniques here, but to gain some insight imagine a squiggly input-output curve of a non-linear system. If we zoom in to any one part of that squiggly line, we can make it look mostly linear. So within some range of inputs we can approximate the system as being linear. We could then approximate some other range of inputs with another line, and so on. The phrase engineers would use here is *piecewise linear*. We now have a number of linear equations that represent the non-linear system.

In this text we will assume that all of our systems are linear.

2.11   STATIONARY AND NON-STATIONARY

In infinite systems (or finite but very complex systems) we discussed the idea of having a random process. But not all random processes are equal. There are different *distributions* for random processes such as Poisson, Gaussian or Boltzmann. And for each there are variables that might change, such as the mean, standard deviation or an exponent. In a *stationary* system the distribution and parameters of that distribution stay the same over time. In a *non-stationary* system they do not. The difference between time-varying and invariant systems is whether the rules change in some way. Stationary and non-stationary systems are similar but it is the nature of the random process that is changing over time.

To gain a bit of insight, imagine that some part of a system is governed by a simple coin flip. The odds are 50/50 and in a stationary system the coin flip will always remain 50/50. We could even imagine a stationary system where the odds are 75/25. What makes a system stationary is that the odds do not change. To make this random coin flip non-stationary, the odds would need to change over time.

Many biological systems are non-stationary. But like linearity, there are degrees of stationarity. Over the course of seconds, minutes, hours or perhaps years, the random processes do not change significantly. There are techniques for dealing with non-stationary systems that are similar in nature to dealing with non-linear systems—for example a researcher might define an *epoch* of time during which it is assumed that the system does not change, and then treat a non-stationary system as many piecewise stationary systems.

In this text we will only consider stationary systems.

2.12   MEMORY AND MEMORILESS SYSTEMS

When the rules that govern the evolution of system states are examined, there is a difference between systems that depend upon their past history from those that erase their history as they go. The systems that depend upon past states must have a way to store those states in some form of memory. Engineers thus talk about memory and memoriless systems. Consider the difference equation below.

$$x[n] = 2x[n-1] + x[n-2] - 3x[n-5] \tag{2.26}$$

In this discrete system, the value of x at time n is dependent upon where x was in the previous time step, but also back two and five time steps ago. We would say that this system has a memory of five time steps. This is an important property in a system—some systems have short memory and others have long memory. You can think about this practically as how many numbers you need to store to keep going with the process of iterating this equation. The other way to think about this is that there must be something within the system (if this equation is a model of a real system) capable of storing at least five system states. Although it is simplest to see the impact of memory in a discrete system, continuous systems certainly have memory as well, for example the charge stored on a capacitor or the inertia of a mass.

We can also have systems that have no memory. All this means is that their evolution is only dependent upon their last state.

$$x[n] = 5 + f[n] + \sin[2(n)] + x[n-1] \qquad (2.27)$$

Here the system is dependent upon two calculations that are reflexive, $f[n]$ and $\sin[2(n)]$. You can think of these as simple look-up tables. The last term is the previous state. Once the system state $x[n]$ is achieved, it erases (some would say overwrites) the previous state, $x[n-1]$.

2.13   TIME CONSTANTS

All physical systems have states. Those states, whether discrete or not, typically cannot make certain jumps instantaneously—a system cannot teleport matter or charges around a system. For example, if the concentration on one side of a membrane is going to change, it will take the passage of enough molecules to make the change. The restriction is that the diffusion of molecules only proceeds at some rate. A concentration across a membrane cannot change instantaneously. The same could be said about other quantities like the build up of electric charge on a capacitor.

What we are after is that many systems have a characteristic time, known as a *time constant*, that gives an indication of how fast things can change in a system. If we are talking about changing the concentration of Calcium in a cup of water, changes might be on the order of minutes or seconds. But if we are trying to change the concentration of Calcium in an ocean, that is going to be on the order of years, decades or centuries. In each case, there is some time constant that is built into the system that will govern how fast things *can* change. Some systems are easy to change and respond quickly (fast or small time constant). You can think of these systems as having little inertia and short *delay times*. Other systems are hard to change, respond slowly (large delay times, slow or large time constant) and have a great deal of inertia.

Given this interpretation, you can think of all sorts of systems that might have time constant:

- Voltage in an RC circuit

- Opening or closing of an ion channel

- Expression of a gene

- Knee jerk reaction

- Spread of an epidemic

- Heating of the skin during thermal ablation

- How quickly you can pick up 90% of the material in this book

- The half-life of a radioactive element

Although, there are many formal definitions of a time constant that are used in the sciences and engineering, there are other fields that have time constants that govern the rates at which languages evolve and die, cultures expand and decay, and friendships develop and fall apart.

It is often possible to measure a signal from a system (output or internal signal) to estimate the time constant. For example, we could measure how an epidemic spreads throughout a population and this might give us some idea of how other quantities (rumors, products, other viruses) might spread. That is because the time constant of the system is the same.

Many systems are the product of two or more time constants. In fact, each variable (or rule) might have its own time constant. This is why the system might behave differently when growing versus decaying, or over long and short time scales. A city might grow slowly and decay very quickly. Second, some systems have time constants that cause some variable in the system to oscillate.

2.14 CONCLUSION

This chapter has introduced you to the many different ways that engineers classify systems. Some overlap and some do not. Some classifications are special cases of a larger classification. But the reason all were introduced is that the assumptions we make will change how we treat a signal or system. In this text, we are going to make some very specific assumptions—our systems will be causal, finite, dynamic, deterministic, time-invariant and linear. We will make these assumptions so that we can use some very powerful analysis tools that have been built up around *systems theory*. We are now ready to start our exploration of these concepts and tools.

2.15 EXERCISES

1. List two examples of heuristics that you use to get you through your day.

2. Which systems below are non-linear and why? You can assume that y and u are independent variables.

$$4\frac{dy}{dt} + y + 4 = \frac{du}{dt} = 5u \qquad (2.28)$$

$$4\frac{d^4y}{dt^4} + a(t)\frac{dy}{dt} = F(t) \qquad (2.29)$$

$$b\frac{d^2y}{dt^2} + \frac{du}{dt} + \sin(35) = \frac{d^3y}{dt^3} \qquad (2.30)$$

$$c(t)\frac{du}{dt} + tan^{-1}(u) = G(t) \tag{2.31}$$

$$d\frac{du}{dt} + \frac{dy}{dt}\frac{du}{dt} = H(t) \tag{2.32}$$

3. For the equations above, what is the order of the system?

4. Which systems above are time-varying?

5. Find three examples of non-stationary systems that you encounter every day. Explain why they are non-stationary.

6. List two systems that you assume to be static and three dynamic systems that you can find in a student cafeteria.

7. Give two examples of a system that has memory. What in the system allows it to have that memory? In other words, what physical quantity is doing the storing?

8. Give three examples of biomedical or biological time constants. Estimate the duration of time over which each system changes.

9. Give two examples of systems that are stable, but have a dynamic equilibrium.

10. List two examples of time-varying biological systems.

CHAPTER 3

System Models

3.1 WHAT IS A MODEL

A model is a representation of a real phenomenon. Like our initial definition of a system, this is not very specific, but will become more specific in this chapter. Consider that you may have played with model airplanes or dolls as a kid. Both are excellent physical models of something—a real airplane or real baby. As such, we don't expect a model airplane or doll to take on all of the characteristics of the real system. A good model will capture only the important aspects of the real phenomenon. For this reason, there are many forms a model can take and there may be many perfectly fine ways to model the same system. Some airplane models may be physical representations but not be capable of flight, while others fly like a plane but look nothing like one. It is good to keep in mind a quote from George Box, "All models are wrong, some happen to be useful." It is the useful models that we will focus on in this chapter.

A major difference between *hard science* and *soft science* is that in a hard science the central ideas of the field have been expressed in mathematical models. Soft sciences on the other hand have more heuristic rules or statistical descriptions. This is certainly not to diminish these fields! In fact, these fields may simply be waiting for an Isaac Newton, Carnot, Boyle or *you* to formulate a mathematical model. Many areas of biology have already crossed over into a hard science while others are on the threshold.

The models we will consider attempt to capture the dynamic properties of a system. The rules that govern these systems will take the form of differential (for continuous systems) or difference equations (for discrete systems). Both provide precise rules for predicting what the system will do next. They are not the only way to model a dynamic system (e.g., agent-based models) and there are advantages to using these alternative methods to capture phenomena that might be missed by the standard modeling methods.

The focus of this chapter is on two methods for generating differential equations that model a real system. The first is to use conservation laws. The second is to use what are known as compartment (or state) models. They are very much related, and in fact, many models will use a bit of both approaches.

3.2 MODELS USING CONSERVATION

Many dynamic models are created based upon conservation laws such as conservation of energy, mass, charge, or momentum. For lack of a better term, we will call the quantities conserved,

"Stuff." It should be noted that "Stuff" can be just about anything, even non-physical stuff. For example, in making an economic model, we might call "Stuff" Gross National Product.

The typical method of quantifying a system with a conservation law is to describe the change in "Stuff" as being equal to the sum of *stuff in* minus the sum of *stuff out* plus any *stuff generated*.

$$\frac{d(Stuff)}{dt} = \sum stuff\,in - \sum stuff\,out + \sum stuff\,generated \tag{3.1}$$

Here *stuff in* is anything that comes into the system and adds to "Stuff." Likewise, *stuff out* is anything that leaves the system and subtracts from "Stuff." For many systems this will be enough. However, there are some systems which have some internal way of generating "Stuff" and this requires an extra term in our model (*stuff generated*). Figure 3.1 is a graphical representation of Equation 3.1.

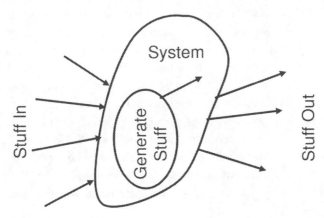

Figure 3.1: A general system showing inputs, outputs and any internal generation of "stuff."

In each discipline *stuff in*, *stuff out*, *stuff generated* may have different names, but in reality they will have the same mathematical form. In the following sections you will see how a number of these fields use conservation principles to write down differential equations that model the system. Something to keep in mind is that the definition for stuff in, stuff out and stuff generated will change if the system is defined in a different way.

Before we jump in, there is a connection to be made. "Stuff" will always have some sort of units—voltage, distance, money, concentration. So the units of $\frac{dStuff}{dt}$ are going to be Stuff/Time.

A bit of thought will show that this is a flow of Stuff. So in Equation 3.1 the terms on the right hand side are flows. Another way to think about this is how fast Stuff is flowing into, out of, or generated by the system. The most common way to create these terms then is

$$Flow\ In = \frac{Stuff}{k_{in}} \tag{3.2}$$

where k_{in} is known by many names that include *time constant* and *rate constant*. In the homework you will be asked to think about the units of this term.

3.2.1 CONSERVATION OF MOMENTUM

To derive mechanical models the conservation property most often used is a balance of forces. You have most likely seen a force balance for a static object like a beam or a section of bone. But $F = ma$ (and other such formulas) work just as well if things are moving too.

Figure 3.2 shows a very simplified schematic of a mechanical air table, the type used in molecular and optical experiments to dampen out any vibrations in the building.

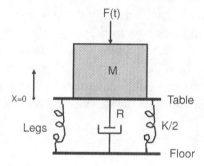

Figure 3.2: A simple mechanical model of an air table.

Here K is a spring constant. The constant has been divided by 2 (one for each leg) just to make the combined spring constant come out to be K. In a real table we would have 4 legs and then just divide by 4. R is a mechanical damper and M is the mass of some weight on the table. The force is pointing down and we are going to define the positive x direction as pointing up. This is because as we apply $F(t)$ downward, all three elements (M, R and K) will provide a force pointing upward.

We can then find the upward forces that each element will provide

$$F_k = \frac{K}{2}x \tag{3.3}$$

$$F_r = R\frac{dx}{dt} \tag{3.4}$$

$$F_m = M\frac{d^2x}{dt^2} \tag{3.5}$$

and we know that the sum of these forces must be equal to $F(t)$

$$F(t) = M\frac{d^2x}{dt^2} + R\frac{dx}{dt} + Kx \tag{3.6}$$

Equation 3.6 is the governing second-order differential equation that governs the dynamics of the force table.

3.2.2   CONSERVATION OF CHARGE

Heldt et al.'s 2002 model (*J. App Physio*) of pressures and flows throughout the circulatory system is composed of a number of repeating tubes that represent the vasculature. But the model does something that you will see in many biomedical models. It models a system in one domain (fluid flow) using models from another domain (electric circuits). This is a very powerful approach because tools from several disciplines can then be used where they are most appropriate. This works because, at some deep level, the elements and forms of the equations are the same. For example, the final form of Equation 3.6 is just a generic second-order equation. The same form (with different constants) could have been derived from an electric circuit or a thermal system.

For historical reasons, often models are translated into the electrical domain. The Heldt et al. model is composed of many copies of the same circuit that can be seen in Figure 3.3.

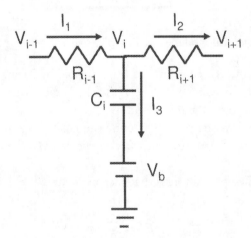

Figure 3.3: A circuit representation of the basic unit used to create the Heldt et al. 2002 model of the circulatory system.

Here voltages represent pressures, currents represent blood flows, capacitors represent reservoirs and resistors are just a constriction. We can use Kirchhoff's Current Law (derived from conservation of charge) to analyze node i. The law says that all currents flowing into and out of that node must be equal.

$$I_1 = \frac{V_{i-1} - V_i}{R_i} \qquad (3.7)$$

$$I_2 = \frac{V_i - V_{i+1}}{R_{i+1}} \qquad (3.8)$$

$$I_3 = \frac{d}{dt}[C_i(V_i - V_b)] \qquad (3.9)$$

and using Kirchhoff's Law

$$I_1 = I_2 + I_3 \qquad (3.10)$$

$$\frac{V_{i-1} - V_i}{R_i} = \frac{V_i - V_{i+1}}{R_{i+1}} + \frac{d}{dt}[C_i(V_i - V_b)] \qquad (3.11)$$

Equation 3.11 is a first-order differential equation for the circuit in Figure 3.3. The full Heldt et al. model is therefore just a series of differential equations of the same form.

3.2.3 CONSERVATION OF MASS

We often think about maintaining our weight at some desired value. Using the conservation of mass, we can take into account mass in and mass out as shown in Figure 3.4. For inputs you typically will take in food (M_f) and liquid (M_l). For outputs you may consider urine (M_u), feces (M_p) and sweat (M_s).

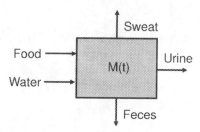

Figure 3.4: A very simple model of mass of the body.

$$\frac{dM}{dt} = aM_f + bM_w - cM_u - dM_p - eM_s \qquad (3.12)$$

The term M is your mass and the terms $a - e$ are constants with units $1/time$ to keep both sides of the equations with equivalent units. You can think of these terms as the rate at which mass is gained or lost (a flow of mass). Using this equation we can consider two situations. First, let's consider that you go to the gym and get a good sweat going for an hour. Since the amount of sweat leaving is positive, the term $-e\,M_s$ is negative. The effect is that $\frac{dM}{dt}$ is negative and you lose some weight. After your workout, you may have a sports drink which adds to the term $b\,M_w$ and $\frac{dM}{dt}$ is positive.

Second, let's consider what happens when you go to a college party. At the party, the terms $a\,M_f$ and $b\,M_w$ are positive (maybe $b\,M_w$ is much greater than $a\,M_f$). The effect is that at the end of the party you have gained some mass. However, later in the night it is possible that the terms $c\,M_u$ and $d\,M_p$ rise, causing $\frac{dM}{dt}$ to become negative. The net effect is that in the morning, you may weigh about the same as before the party (although you might be dehydrated).

3.2.4   FLUID MASS AND VOLUME

Consider the liquid tank in Figure 3.5 with the inflow, q_i, and outflow, q_o. We can first find the mass of the liquid in the tank. Here we can define the density of the liquid as ρ.

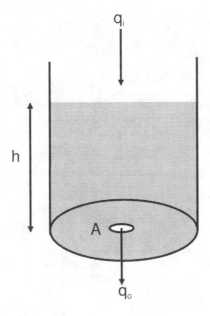

Figure 3.5: A simple unsteady tank problem.

Then the mass in the tank is

$$m = \rho V = \rho A h \tag{3.13}$$

where h is the height of the fluid and A is the cross sectional area. If we want to describe how the mass in the tank varies over time we can express all masses as flow rates (q-terms) times the density, or $q\rho$. Therefore the conservation law tells us that

$$\frac{dm}{dt} = \rho q_i - \rho q_o \tag{3.14}$$

where the inflow and outflow are in units of liters per second.

From this model we may be more interested in some other value that changes over time, for example the height, h, of the liquid in the tank. Here we need to make a relationship between m and h using the volume of fluid in the tank.

$$m = \rho V = \rho A h \tag{3.15}$$

We can substitute this relationship into Equation 3.14 to get

$$\frac{dm}{dt} = \frac{d(\rho V)}{dt} = \frac{d(\rho A h)}{dt} = \rho A \frac{dh}{dt} \tag{3.16}$$

$$\rho A \frac{dh}{dt} = \rho q_i - \rho q_o \tag{3.17}$$

Or more simply,

$$\frac{dh}{dt} = \frac{1}{A}(q_i - q_o) \tag{3.18}$$

In the context of signals and systems, we have now derived an equation that will tell us a measurable output, h, given flows in and out of the system. This is a great example of how we can manipulate equations to redefine what we mean by inputs and outputs.

3.2.5 CONSERVATION OF ENERGY

One of the most powerful conservation laws tells us that energy cannot be created or destroyed. So the total energy of a system, E_{total}, can only be changed if energy either enters, E_i, or exits, E_o, the system

$$\frac{dE_{total}}{dt} = E_i - E_o \tag{3.19}$$

A common assumption in many systems is that there is no energy entering or exiting the system and therefore, $\frac{dE_{total}}{dt} = 0$. On the surface this may lead you to believe that nothing interesting can happen. The tricky thing about energy, however, is that it can take on many forms: kinetic, potential, thermal, pressure and others. The key here is that while the total energy may stay the same, the form of energy can change types. So in reality

$$E_{total} = E_{potential} + E_{kinetic} + E_{thermal}\cdots \tag{3.20}$$

We can extend this idea to derive Bernoulli's equation by starting with

$$E_{total} = E_{pressure} + E_{potential} + E_{Kinetic} \tag{3.21}$$

Substitution of some expressions for the energy terms for fluid flow gives us

$$E_{total} = P + \rho gh + \frac{1}{2}\rho v^2 \tag{3.22}$$

where P is the pressure, ρ is the density of the fluid, g is gravity, h is height, and v is the velocity of the fluid. Now we can consider a vein or artery in the body as shown in Figure 3.6.

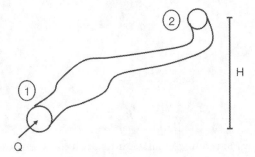

Figure 3.6: An example of a vein or artery in the body.

In this system we can assume that the total energy of the system is not changing ($\frac{dE_{total}}{dt} = 0$), from point 1 to point 2 even though the height, pressure and artery diameter change. So the energy at point 1 must be equal to the energy at point 2.

$$E_1 = E_2 \tag{3.23}$$
$$E_{pressure1} + E_{potential1} + E_{kinetic1} = E_{pressure2} + E_{potential2} + E_{kinetic2} \tag{3.24}$$

or more specifically

$$P_1 + \rho gh_1 + \frac{1}{2}\rho v_1^2 = P_2 + \rho gh_2 + \frac{1}{2}\rho v_2^2 \tag{3.25}$$

With a little rearranging

$$(P_2 - P_1) + \rho g(h_2 - h_1) + \frac{1}{2}\rho(v_2^2 - v_1^2) = 0 \tag{3.26}$$

$$\Delta P + \rho g \Delta h + \frac{1}{2}\rho \Delta(v^2) = 0 \tag{3.27}$$

Using this equation we can rearrange to find any of the changes (Δ 's) that occur as a result of some other change.

We have neglected two important terms to make our energy balance—friction and any generator of flow. The friction is a loss of energy from the system, usually as heat. The generator is usually some sort of pump. In a biological context the most obvious pump is the heart. But arteries and arterioles are lined with smooth muscle which in some cases may help move blood along. We will not write these out in specific terms but rather show where they appear in the equation.

$$\Delta P + \rho g \Delta h + \frac{1}{2}\rho \Delta(v^2) + E_{pump} - E_{friction} = 0 \tag{3.28}$$

Using the idea of conservation of energy we now have an equation that governs how different types of energy are transformed into one another as blood flows through the body.

3.2.6 OTHER MODELS

There are certainly many other ways to use conservation laws to derive equations for bioacoustic, biothermal and biooptical systems. We can even extend the ideas of conservation well past the usual physical models considered by engineers. For example, there are a number of conservation models in the social sciences and even in the arts and humanities.

Consider two competitive companies engaged in advertising campaigns. It is reasonable to assume that the rate at which one company increases its sales is proportional to the available market. The constants of proportionality will depend on the effectiveness of each campaign. Thus we can develop the following differential equations

$$\frac{dS_1}{dt} = C_1 M_a \tag{3.29}$$

$$\frac{dS_2}{dt} = C_2 M_a \tag{3.30}$$

where $S_i(t)$ is the sales of the company i in any year (sales/year) and C_1 and C_2 are constants. $M_a(t)$ is the available market.

$$M_a = M(t) - S_1(t) - S_2(t) \tag{3.31}$$

where $M(t)$ is the total market. With a bit of thought we might even be able to derive an equation that governs how the size of the market changes, $\frac{dM_a}{dt}$.

Using this model we can consider a number of business questions: 1) Suppose you are thinking of starting up a company—company 3. How much of the total market is still untapped? 2) If you are company 3, how much do you need to increase your advertising to keep pace with company 1? Will it be profitable? 3) If you are company 2, is it worth it to buy out company 1? How many years will it take for this move to be profitable? The power of the model in this case is that you can get some real quantitative predictions to these questions that can help you make business decisions.

A socio-political example is the Lanchester War Model. Here we suppose that one side has N_1 units, each with a hitting power α. They are engaged with a second side, having N_2 units, each with hitting power β. Suppose further that the engagement is such that the fire power is directed equally against all opponents and vice versa (i.e., a two-sided war). The rate of loss of the two forces can be written as:

$$\frac{dN_1}{dt} = -k_2\beta N_2 \tag{3.32}$$

$$\frac{dN_2}{dt} = -k_1\alpha N_1 \tag{3.33}$$

where k_1 and k_2 are positive constants, N_1 and N_2 are in fact integers (numbers of soldiers) but we can treat them as continuous variables for this example. The strength of the two forces is defined as equal when their fractional losses are equal, that is, when

$$\frac{1}{N_1}\frac{dN_1}{dt} = \frac{1}{N_2}\frac{dN_2}{dt} \tag{3.34}$$

This type of analysis is used in balance of power problems of all kinds (e.g., electrical power, fluid flow). On dividing the first equation by the second and integrating, we obtain two hyperbolas, defined by $\alpha N_1^2 - \beta N_2^2 = C$. Taking $C = 0$ gives Lanchester's N^2 law, which states that the strength of a force is proportional to the fire power of a unit multiplied by the square of the number of units.

The point of the Lanchester War Model and the Advertising model is not for you to memorize them, but rather to understand that other fields take the same systems approach as us.

3.3   STATE AND COMPARTMENT MODELS

Another way in which differential equations are used to build a mathematical model is to assume that the system contains either compartments that trade some quantity, or that the system has several states. In both cases, the model consists of several *coupled* differential equations, one for each state or compartment.

3.3.1   VOLUME BALANCE

Liquid in the body can be considered to be shared between three different compartments at shown in Figure 3.7: the gut, blood and bladder.

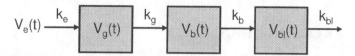

Figure 3.7: A model of fluid flow from the gut to the blood to the bladder.

Fluid enters the gut from the outside and moves through the intestines where it diffuses across a membrane into the blood. So the input to the gut is a volume of fluid from the environment (V_e) at a certain rate. We can assume that this volume is entering at some constant rate (clearly an approximation), k_e. So the rate at which fluid is entering the body is a term with the units volume/time.

$$k_e V_e \qquad (3.35)$$

We also have some amount of volume in the gut already, V_g. But it is leaking out to the intestines (and then to the blood) at some rate, k_g. Given all of the terms we have defined, we can write down an equation to describe how the volume of fluid in the gut will change.

$$\frac{dV_g}{dt} = k_e V_e - k_g V_g \qquad (3.36)$$

What we are building up is a very powerful approach for handling all types of *compartment* models, those where there is a flow between compartments. Our next step is to consider the blood. We can use logic similar to how we derived the differential equation for the gut, to write down an equation for the blood

$$\frac{dV_b}{dt} = k_g V_g - k_b V_b \qquad (3.37)$$

Notice how the term $k_g V_g$ appears in both compartments, but it is the output of the gut (subtracted from V_g) and input to the blood (added to V_b). Lastly we can write down the equation for the bladder

$$\frac{dV_b}{dt} = k_b V_b - k_{bl} V_{bl} \qquad (3.38)$$

In the end, we have three coupled differential equations that describe how fluid enters and finally leaves the body. There are several flaws in this model. For example, fluid could be absorbed in cells and cause you to be bloated. We have also neglected the fact that volumes of fluid do not suddenly cross a membrane but do so slowly through diffusion or some active process. But we can easily account for these by making the movement a bit more complex (e.g., using Fick's Law of Diffusion).

3.3.2   MODELS OF ION CHANNELS

One of the more famous mathematical models is the Hodgkin-Huxley model of action potentials in a giant squid neuron. To accomplish this modeling feat (for which they won the Nobel Prize in Medicine in 1962) Hodgkin and Huxley assumed that an ion channel could be in either of two states: open or closed. When a channel was open, ionic current would flow across the cell membrane. When the channel was closed, ionic current would not flow. So they didn't need to consider every ion channel, but instead considered the probability of a channel being open or closed. Furthermore, they assumed that the rate at which channels could transition from one state to the next were dependent upon the voltage across the membrane (V_m) of the cell. Given these assumptions they formulated the most basic channel dynamics as:

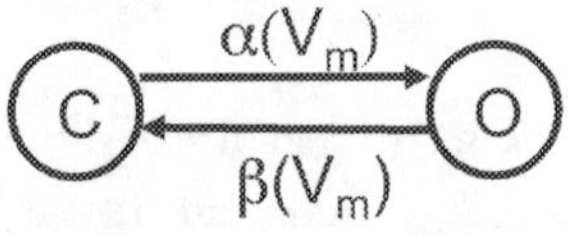

Figure 3.8: A generic schematic of the open and closed states of an ion channel.

$$\frac{dO}{dt} = \alpha(V_m)O - \beta(V_m)(1 - O) \tag{3.39}$$

where O is the percentage of Potassium channels that are open. They then went on to find functions for $\alpha(V_m)$ and $\beta(V_m)$ from their experiments. Again, we can think of this as a system. The change in probability of a channel being open is simply how many channels are opening minus how many channels are closing.

This approach was very useful for modeling ion channels. But, as technology progressed, it became possible to consider more than just open and closed states. A recently published model of the Potassium channels is shown in Figure 3.9.

This is an example of what is known as a Markov Model—a system can change from one state to another in a fluid way. Here we have one open state (O), but a number of other states that the ion channel might be in. These are analogous to the proteins that make up the channel twisting and turning in different configurations. Some configurations leave the channel *open*. In other configurations, the channel is *closed* and not showing any signs of being open. But there are some configurations where all it takes is for a few segments of the protein to move for the channel to be open. These configurations are called *inactivated*.

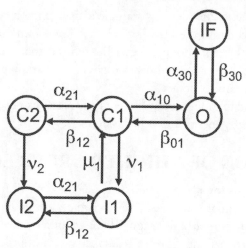

Figure 3.9: A relatively complex state diagram for a Potassium ion channel.

Here we can apply the same idea to derive a set of coupled differential equations:

$$\frac{dIF}{dt} = \alpha_{30}O - \beta_{30}IF \tag{3.40}$$

$$\frac{dO}{dt} = \alpha_{10}C1 + \beta_{30}IF - \beta_{01}O - \alpha_{30}O \tag{3.41}$$

$$\frac{dC1}{dt} = \alpha_{21}C2 + \beta_{01}O + \mu_1 I1 - \beta_{12}C1 - \alpha_{10}C1 - v_1 C1 \tag{3.42}$$

$$\frac{dC2}{dt} = \beta_{12}C1 - \alpha_{21}C2 - v_2 C2 \tag{3.43}$$

$$\frac{dI1}{dt} = \alpha_{21}I2 + v_1 C1 - \beta_{12}I1 - \mu_1 I1 \tag{3.44}$$

$$\frac{dI2}{dt} = v_2 C2 + \beta_{12}I1 - \alpha_{21}I2 \tag{3.45}$$

The general pattern in each equation is to sum up all flows in and subtract all outgoing flows. All terms on the right side are then a value of a state multiplied by a rate constant. So the combined terms are flows (and have the right units). The one catch is that when considering a particular state, all inflows are multiplied by the state that is contributing to the inflow. Outflows, on the other hand, are multiplied by their own state. A good rule of thumb is that you multiply by the compartment of origin.

One last point is that the equations above are all linear. That means that they are subject to the linear analysis that we will build up in the chapters to come.

So why make a Potassium channel model this complicated? Why not just use the original Hodgkin-Huxely model? There are two reasons: First, a mutation in a single protein may only impact one part of the Markov model, for example a change in one of the k terms or a deletion or addition of a state. Second, in the Hodgkin-Huxely Potassium model, the α and β terms are dependent upon V_m and therefore not constant. The result is that the differential equation is not linear and we cannot apply our system tools as easily. The Markov model, on the other hand, is linear because the k terms are constants.

The same ideas used to model compartments and states can be used to derive equations for flows in ecological, business, energy and other systems.

3.4 REDUCTION OF A HIGHER ORDER EQUATION

Often differential equations for a large system (such as the Potassium channel above) are written as several first-order differential equations. There are many advantages of doing this, for example we could write the system in matrix notation. There are other times that the derivation yields a single higher-order differential equation. In reality, a system of order n can be represented as n first-order differential equations, or a single equation with an nth-order derivative.

It is possible to convert back and forth between the two with a little work. As a simple example, remember the term from the air table example, $F_m = M \frac{d^2 x}{dt^2}$. This is a second-order term. But we can turn it into two first-order terms. First, we can define

$$v = \frac{dx}{dt} \tag{3.46}$$

Then through substitution

$$F_m = \frac{dv}{dt} \tag{3.47}$$

and so we have the two very simple first-order differential equations

$$\frac{dx}{dt} = v \tag{3.48}$$

$$\frac{dv}{dt} = F_m \tag{3.49}$$

that are equivalent to the higher-order term. We can do the same for any arbitrarily high-order term.

3.5 EXERCISES

1. The passive membrane of any cell can be modeled (Figure 3.10) as a capacitor (C_m) in parallel with a resistor (R_m) and battery (V_{rest}). The voltage across the membrane is simply

the difference between the potential outside the cell (0V or ground) and the potential inside the cell (V_m). Assume that current is flowing in the intracellular space from cell 1 to cell 2 to cell 3. The current flows from cell to cell through gap junctions that may be modeled as resistors (R_i).

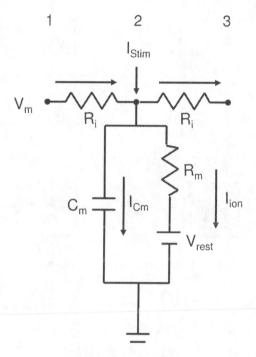

Figure 3.10: Passive circuit model for a cell membrane model.

    a. Write an equation for the capacitive current (I_{Cm}) at node 2.

    b. Write an equation for the ionic current (I_{ion}) at node 2.

    c. Derive an equation to describe how Vm at node 2 changes over time (e.g., $\frac{dV_m}{dt} =$). Your expression should contain only voltages, C_m, R_m, R_i, and I_{stim}.

2. An excised frog muscle is suspended from a support and pre-stretched by a weight (Figure 3.11). Consider this situation as a system with the input being an additional applied force and the output being the position of the weight, x. The passive (i.e., not contracting) muscle can be represented as a spring in parallel with a viscous resistance (the Voight muscle model). Write a differential equation for the position, x, as a function of time.

3. Kalant et al., in *Biochemical Pharmacology* (24:431–434) published a model (Figure 3.12) of how the body handles the absorption and elimination of alcohol. Below is a modified compartment model that includes the concentrations (mM) in the gut, blood and kidney.

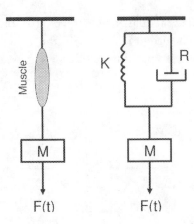

Figure 3.11: Voight model of a frog muscle supported and stretched by a weight.

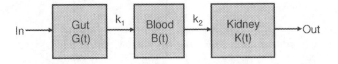

Figure 3.12: A simple model of flow from the gut to blood to kidney.

    a. Write down the governing differential equations for this system.

    b. What are the units of k1 and k2?

    c. What are the units of the input and output?

4. Write down a series of first-order differential equations for the following wild-type Sodium channel shown in Figure 3.13.

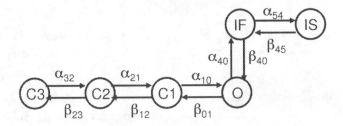

Figure 3.13: A schematic of the state transitions in a Sodium ion channel.

CHAPTER 4

Laplace Transform

4.1 INTRODUCTION

The Laplace Transform has a mathematical definition, but don't let that get in the way of its usefulness in solving engineering problems. At its heart, the Laplace Transform is often the easiest pathway toward solving a linear differential equation. In the context of signals and systems, solving means generating signals from a system that is governed by the differential equation.

Three relatively simple steps are needed to find a time signal from a linear differential equation. In the first step the Laplace Transform converts equations in differential notation (the language of calculus) into an algebraic equation. Second, the algebraic equation can be solved by simple manipulation (addition/subtraction, multiplication/division) to isolate the term that varies over time. Third, the inverse Laplace Transform can be applied to yield the time solution. We can summarize these steps graphically in Figure 4.1 as a *mapping* between the *Time Domain* and *s-domain*.

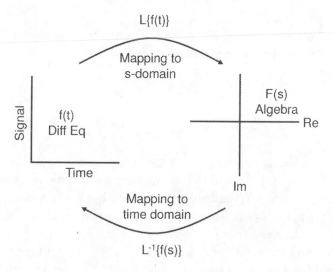

Figure 4.1: Mapping from the Time Domain to Laplace Domain and back again.

In this chapter we will focus on how to make the mappings from the time to the s-domain and then back again. We will also briefly describe the equivalent transform for discrete signals, the z-transform.

4.2   FORMAL DEFINITIONS

Figure 4.2 shows how we might view the Laplace Transform as a system that operates on inputs and outputs. We will focus primarily on the forward (time to s-domain direction) in this chapter.

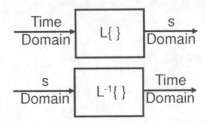

Figure 4.2: A systems block diagram of the forward and inverse Laplace Transform.

4.2.1   LAPLACE TRANSFORM

The mathematical definition of the Laplace Transform is:

$$x(s) = L\left[x(t)\right] = \int_{-\infty}^{\infty} x(t)e^{-st}\,dt \tag{4.1}$$

Equation 4.1 is called the two-sided transform and works for time both in the past the future. However, since engineers typically are not concerned with negative time, we use the *one-sided* Laplace Transform:

$$x(s) = L\left[x(t)\right] = \int_{0}^{\infty} x(t)e^{-st}\,dt \tag{4.2}$$

Where L indicates the Laplace operation. Another way to think about this is that from negative infinity to zero the signal $x(t) = 0$ and so does not contribute to the integral.

It is important to note that the Laplace Transform is a linear operation and will therefore obey all of the properties of linear systems. This is important because if x(t) is a signal from a linear system, then the Laplace Transform will preserve the linearity.

Of note is that on the right hand side we have a function in time, $x(t)$, and on the left hand side we have a function of s. One way to think of this is that the behavior of your system in time has been *mapped* (or transformed) into the s-domain, as in Figure 4.1. A *mapping* is any transformation where a function of one variable, $x(t)$, has been transformed into a function of another variable, $x(s)$. A similar engineering concept is when you need to change units or when one energy is transformed from one type to another by a transducer (see section 1.2.2). Ideally no information is lost in the transformation.

The term s in the Laplace Transform is a complex number, often represented as $\sigma + j\omega$. In future chapters we will visualize systems and signals by plotting them onto a σ-ω, or real-imaginary axis. These plots will be a graphical representation of the system that will allow us to assess properties such as stability.

The Laplace Transform is one of a series of transforms that are used in math, physics and engineering. In fact, you will encounter the Fourier Transform in Chapter 11. What is important to know about transforms is that they move information from one *domain* into another *domain*, similar to moving text from one language to another. In some cases no information is lost in the translation, it is simply packaged differently, often with the intent of bringing to the foreground something that was hard to see. In other cases, information is lost in the translation.

4.2.2 INVERSE LAPLACE TRANSFORM

Many transforms are valuable because they work in a forward direction, but can also be reversed. In other words, you can undo a transform with its *inverse*, moving in Figure 4.2 from the s to the Time Domain. The inverse Laplace Transform is defined as:

$$x(t) = L^{-1}\left[x(s)\right] = \frac{1}{2\pi j}\int_{\sigma-j\infty}^{\sigma+j\infty} x(t)e^{st}\,dt \tag{4.3}$$

Where L^{-1} indicates the inverse of the Laplace Transform. You will note that here the limits of integration are not as intuitive. This again is because we have mapped the time variable to the s variable. Typically, $x(s)$ is some polynomial that has roots (also known as *poles*). We will explore this idea further in Chapter 6. For those who do not like this definition, don't worry. We will not take inverses using Equation 4.3.

4.3 TRANSFORM TABLES

Transform tables are created by applying Equations 4.2 and 4.3 to common signals (see Appendix C). The purpose of such a look-up table is so that you don't need to rederive the Laplace Transform every time you want to use it. And because L is a linear operator, you can use the principles of superposition to any of the relationships in the table to build up transforms for more complex signals.

Another helpful property of the Laplace Transform is that it is *symmetric*. That means that if a result is derived, mapping from time to s, the same result will hold simply by switching the t and s variables. For example, in the transform table in Appendix C you will find the following property

$$L\left[f(t - t_0)\right] = f(s)e^{-st_0} \tag{4.4}$$

meaning that shifting a function in time will result in the same Laplace Transform but multiplied by an exponential. What symmetry means is that we can flip this relationship around.

$$L\left[e^{-at} f(t)\right] = f(s + a) \tag{4.5}$$

This is useful in using a transform table, because you actually get twice as many transform pairs as are listed.

4.4 FOUR USEFUL LAPLACE TRANSFORMS

There are three simple signals that we will be using over and over again in the coming chapters. They are the *impulse*, *step* and *sinusoid* and we will represent them using functions. There is an additional operator that will be very useful and that is the *derivative*. Below we explain each and show the derivation of their respective Laplace Transforms.

4.4.1 THE IMPULSE

An impulse, $\delta(t)$, is defined as a very brief jump from 0 to ∞ and then back to 0 again:

$$\delta(t) = \begin{cases} \infty & \text{if } t = 0 \\ 0 & \text{Otherwise} \end{cases}$$

but, the area under this impulse is defined to be one.

$$\int_{-\infty}^{\infty} \delta(t)dt = 1 \tag{4.6}$$

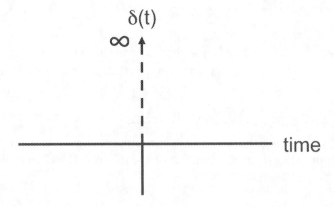

Figure 4.3: Impulse function ($\delta(t)$) that takes on a value of one only for time zero.

$\delta(t)$ is sometimes called the *Dirac Delta Function*, shown in Figure 4.3. To derive the Laplace Transform we can simply plug in Equation 4.2.

$$L\left[\delta(t)\right] = \int_{-\infty}^{\infty} \delta(t)e^{-st}\,dt \tag{4.7}$$

The catch here is that $\delta(t)$ will only contribute to the integral at $t = 0$. To make sense of how to take this integral, we will instead think of δ contributing between $-\epsilon$ and $+\epsilon$, where ϵ is a vanishingly small number. But the catch is that in that interval, we know that the function is infinity.

$$L\left[\delta(t)\right] = \int_{-\epsilon}^{\epsilon} \delta(t)e^{-st}\,dt \tag{4.8}$$

Through some applications of l'Hospital's rule we can find that

$$L\left[\delta(t)\right] = 1 \tag{4.9}$$

A more intuitive, but more sloppy, derivation is to recognize that when $t = 0$ (which is the only place where $\delta(t)$ has any value) the exponential (e^{-st}) is also equal to 1.

$$L\left[\delta(t)\right] = \int_{0-}^{0+} \delta(t)1\,dt \tag{4.10}$$

$$\tag{4.11}$$

and using Equation 4.6

$$L\left[\delta(t)\right] = 1 \tag{4.12}$$

4.4.2 THE UNIT STEP

A step function is defined as:

$$u(t) = \begin{cases} 1 & \text{if } t > 0 \\ 0 & \text{Otherwise} \end{cases}$$

$u(t)$ is also sometimes just called the *step* function or *Heaviside* function. Graphically, the step function is shown in Figure 4.4.

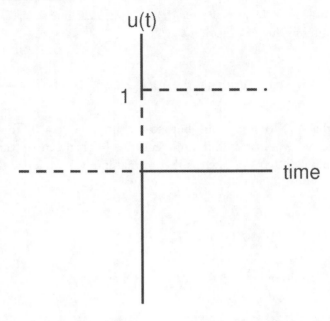

Figure 4.4: Step function ($u(t)$) is zero before $t = 0$ and one after $t = 0$.

Substitution yields

$$L\left[u(t)\right] = \int_{-\infty}^{\infty} u(t)e^{-st}\,dt \qquad (4.13)$$

but we know that for $t < 0$, $u(t) = 0$ and therefore does not contribute to the integral. For $t > 0$, $u(t) = 1$. So,

$$L\left[u(t)\right] = \int_{0}^{\infty} 1e^{-st}\,dt \qquad (4.14)$$

$$L\left[u(t)\right] = \int_{0}^{\infty} e^{-st}\,dt \qquad (4.15)$$

then

$$L[u(t)] = \int_0^{\infty} e^{-st} dt \tag{4.16}$$

$$= -\frac{1}{s} \left[e^{-st} \right]_0^{\infty} \tag{4.17}$$

$$= -\frac{1}{s} [0 - 1] \tag{4.18}$$

$$= \frac{1}{s} \tag{4.19}$$

It also helps to think of the graphical interpretation as the area under a decaying exponential, where s controls the rate of decay.

4.4.3   THE SINUSOID

We can mathematically express a sine wave in two different ways:

$$x(t) = A\sin(\omega t + \theta) \tag{4.20}$$

and graphically in Figure 4.5 where A is the amplitude, ω is the frequency in rad/s, and θ is the

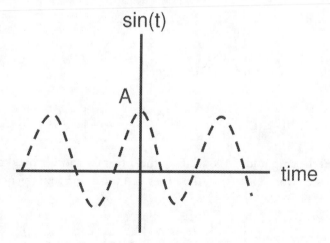

Figure 4.5: A sinusoidal function that oscillates between a minimum and maximum value (A) at a particular frequency (ω), and with a phase shift in time, θ

phase in rad. We need only these three parameters to specify any pure sinusoid. A common and more intuitive way to write the same equation is as

$$x(t) = A\sin(2\pi f t + \theta) \tag{4.21}$$

where f is a frequency in Hz $(1/s)$ and θ is now in the units of degrees. Note that converting between rad/s and Hz can be accomplished by

$$2\pi f = \omega \tag{4.22}$$

The second common way to write a sinusoid, and the way that will be used in this text, is

$$x(t) = Ae^{j\omega t} \tag{4.23}$$

Here A is again the amplitude and $j\omega$ is a complex number that represents both the frequency (real part) and phase (imaginary part). You can see Appendix A for more information on the interpretation of Equation 4.23. One reason this complex formulation for a sinusoid is useful is because of the ease of taking derivatives and Laplace Transforms of exponentials.

For example, when we try to take the Laplace Transform, it is much easier to deal with the exponential form of the sinusoid. But, as shown in Appendix A, there is a difference between a generic sinusoid and the more specific sine signal.

$$\sin(\omega t) = \frac{e^{j\omega t} - e^{-j\omega t}}{2j} \tag{4.24}$$

So to take the Laplace Transform of the sine signal

$$L\left[A\sin(\omega t)\right] = A\int_{-\infty}^{\infty}\left[\frac{e^{j\omega t} - e^{-j\omega t}}{2j}\right]e^{-st}\,dt \tag{4.25}$$

Note that we pulled the amplitude, A, outside of the integral. Another trick, now that we understand how the unit step works, is to turn the two-sided transform into the one-sided transform.

$$L\left[A\sin(\omega t)\right] = A\int_{-\infty}^{\infty}u(t)\left[\frac{e^{j\omega t} - e^{-j\omega t}}{2j}\right]e^{-st}\,dt \tag{4.26}$$

$$= A\int_{0}^{\infty}\left[\frac{e^{j\omega t} - e^{-j\omega t}}{2j}\right]e^{-st}\,dt \tag{4.27}$$

Note that this trick will work with any function that we would like to be zeroed out before $t = 0$. We can now combine terms together in the integral.

$$L\left[A\sin(\omega t)\right] = \frac{A}{2j} \int_0^\infty \left[e^{j\omega t} - e^{-j\omega t}\right] e^{-st} dt \tag{4.28}$$

And now we need to redistribute some of the exponentials

$$L\left[A\sin(\omega t)\right] = \frac{A}{2j} \int_0^\infty e^{j\omega t - st} - e^{-j\omega t - st} dt \tag{4.29}$$

and then evaluation of the integral

$$L\left[A\sin(\omega t)\right] = \frac{A}{2j} \int_0^\infty e^{j\omega t - st} - e^{-j\omega t - st} dt \tag{4.30}$$

$$= \frac{A}{2j} \left[\frac{e^{-(s-j\omega)t}}{-(s-j\omega)} - \frac{e^{-(s+j\omega t)}}{-(s+j\omega)}\right]_0^\infty \tag{4.31}$$

$$= \frac{A}{2j} \left[\frac{1}{s-j\omega} - \frac{1}{s+j\omega}\right] \tag{4.32}$$

$$= A\frac{\omega}{s^2 + \omega^2} \tag{4.33}$$

4.4.4   THE DERIVATIVE

Perhaps the most valuable property of the Laplace Transform is what it does to the derivative. Imagine that we have a derivative, $\frac{dx}{dt}$, or $\dot{x}$, then

$$L\left[\dot{x}(t)\right] = \int_0^\infty \dot{x}(t)e^{-st} dt \tag{4.34}$$

$$\tag{4.35}$$

Integration by parts yields

$$L\left[\dot{x}(t)\right] = e^{-st}x(t)\Big|_0^\infty - \int_0^\infty x(t)(-se^{-st})dt \tag{4.36}$$

The first term is easy to evaluate

$$e^{-st}x(t)\Big|_0^\infty = -x(0) \tag{4.37}$$

In other words, it is simply the initial condition.

$$L\left[\dot{x}\right] = -x(0) - \int_0^\infty x(t)(-se^{-st})dt \tag{4.38}$$

The second term is a bit trickier but in reality it is the definition of the Laplace Transform of $x(t)$ with the addition of an s term. Because our integral is over time, we can in fact pull out this s-term to yield

$$L\left[\dot{x}\right] = -x(0) + sx(s) \tag{4.39}$$

where $x(s) = L[x(t)]$.

This result can be generalized to derivatives of higher power. For example we can substitute $\dot{x}(t)$ for $x(t)$ into Equation 4.39.

$$L\left[\ddot{x}(t)\right] = -\dot{x}(0) + sL[\dot{x}(t)] \tag{4.40}$$
$$= -\dot{x}(0) + s\left[sx(s) - x(0)\right] \tag{4.41}$$
$$= s^2 x(s) - sx(0) - \dot{x}(0) \tag{4.42}$$

Note that for a second derivative, we have an s^2 term, rather than a s term. This is at the heart of what allows the Laplace Transform to turn a differential equation into a polynomial. The degree (order) of the differential equation will determine the power of s that will be present. You should also note that because this is a second-order differential equation, we need two initial conditions, $x(0)$ and $\dot{x}(0)$.

To get the terms for high derivatives, we can keep substituting. For example,

$$L\left[\frac{d^3 x}{dt^3}\right] = s^3 x(s) - s^2 x(0) - s\dot{x}(0) - \ddot{x}(0) \tag{4.43}$$

Notice that in all terms after $s^3 x(s)$, the powers of s go down, while the degree of initial condition for x goes up. The general formula for the transform of any higher derivative is

$$L\left[\frac{d^n x}{dt^n}\right] = s^n x(s) - \sum_{k=1}^n s^{k-1} f^{n-k}(0) \tag{4.44}$$

What is more, the idea of powers of s reflecting the degree of the derivative even extends to integrals

$$L\left[\int x(t)dt\right] = s^{-1}x(s) = \frac{1}{s}x(s) \tag{4.45}$$

4.5   FROM DIFFERENTIAL TO ALGEBRAIC EQUATIONS

In the introduction it was stated that the main advantage of the Laplace Transform in solving differential equations is that it turns them into algebraic equations. We will now take a look at a generic second-order equation.

$$A\frac{d^2y}{dt^2} + B\frac{dy}{dt} + Cy(t) = 0 \tag{4.46}$$

Where $A - C$ are constants. The goal is to find $y(t)$. We also need to know our initial conditions and since this is a second-order system, we need two. For our purposes, we will assume

$$y(0) = -1 \tag{4.47}$$
$$\dot{y}(0) = \frac{dy}{dt}\big|_{t=0} = 2 \tag{4.48}$$

By applying the relationship we derived for the Laplace Transform of a derivative we find:

$$As^2y(s) - Asy(0) - A\dot{y}(0) + Bsy(s) - By(0) + Cy(s) = 0 \tag{4.49}$$
$$As^2y(s) + As - 2A + Bsy(s) + B + Cy(s) = 0 \tag{4.50}$$
$$As^2y(s) + Bsy(s) + Cy(s) + As - 2A + B = 0 \tag{4.51}$$

We now rearranged this *algebraic* equation to arrive at

$$y(s) = \frac{-As + 2A - B}{As^2 + Bs + C} \tag{4.52}$$

And this is the polynomial representation of the signal $y(t)$ in the $s - domain$.

4.6   FROM ALGEBRAIC EQUATIONS TO A SOLUTION

One possible method of solving Equation 4.52 would be to use the definition of the inverse Laplace Transform (Equation 4.3) or relationships from the table in Appendix C. However, there is a much more simple way to arrive at a solution. We know that the solution to any linear differential equation will be the sum of exponential functions.

$$y(t) = a_n e^{\sigma_n t} + a_{n-1} e^{\sigma_{n-1} t} + \ldots + a_2 e^{\sigma_2 t} + a_1 e^{\sigma_1 t} \tag{4.53}$$

Where the σ terms are the roots of the denominator of Equation 4.52

$$y(s) = \frac{?}{(s - \sigma_n)(s - \sigma_{n-1}) \dots (s - \sigma_2)(s - \sigma_1)} \tag{4.54}$$

The reason for the ? is that we cannot find the constants in Equation 4.53 (a terms) if the equation for $y(s)$ is in the form of Equation 4.54. Instead we can transform our equation to the form:

$$y(s) = \frac{a_n}{(s - \sigma_n)} + \frac{a_{n-1}}{(s - \sigma_{n-1})} + \dots \frac{a_2}{(s - \sigma_2)} + \frac{a_1}{(s - \sigma_1)} \tag{4.55}$$

This may be accomplished using a partial fraction expansion as explained in Appendix B.

For our second-order example, however, we will only have $\sigma_{1,2}$ and may find them using the quadratic equation. A few interesting properties of this generic second-order system are worth considering. First, unlike a first-order system we can have two imaginary roots for $\sigma_{1,2}$. What does this mean for our solution in time? (Hint: Look at Euler's identity in Appendix A). Second, consider how the solution would differ if we included some *forcing function* to our system:

$$A \frac{d^2 y}{dt^2} + B \frac{dy}{dt} + C y(t) = f(t) \tag{4.56}$$

where $f(t)$ could be any generic function. You can think of it as an input. If $f(t)$ is some nice analytic function, such as $u(t)$ or $\delta(t)$, then we can simply take the Laplace Transform along with the left hand side of Equation 4.56. The idea of a forcing function will become important in Chapter 8 where we consider the many possible responses of a system.

4.7   OTHER INTERESTING APPLICATIONS

4.7.1   THE FOURIER TRANSFORM

You will discover in Chapter 11 that the Laplace and Fourier Transforms are very similar. In fact, they are identical if $s = j\omega$. This being the case, you may see that some engineers use the Fourier Transform to solve equations in the same way we have used the Laplace Transform. The difference is that the mapping to the Fourier Domain has a more intuitive interpretation which will become more clear in Chapter 11.

4.7.2   NON-TIME MAPPING

There is no reason why the Laplace Transform needs to be used to solve Time Domain differential equations. In fact, everything will work just fine if the differential equations are functions of space ($\frac{d}{dx}$). This is a powerful statement for two reasons. First, we think of time as always moving in one direction, but we can move forward and backward in space. The two-sided Laplace Transform works just fine for forward or backward motion. Second, there are three dimensions of space.

Again, the Laplace Transform can handle the multiple dimensions. Note that when dealing with space, the initial conditions (used for time equations) are replaced by boundary conditions. Lastly, you may now realize that *any* differential equation, regardless of the independent variable, can be operated on by the Laplace Transform. For example, you can imagine an anthropological model that does not use time or space but some other variable.

Since the Laplace Transform can be used for both differential equations in time and space, it is possible to apply the transform to *partial* differential equations (i.e., differential equations that depend on both time and space). The solution of partial differential equations is outside the scope of this book but you should be aware that such methods do exist.

4.8   THE Z-TRANSFORM

The Laplace Transform is for transforming continuous time signals into continuous s-domain signals. The equivalent transform for discrete signals is called the *z-transform*. It can be derived from Equation 4.2 by sampling in the Time Domain.

$$x[e^s] = \sum_{n=0}^{\infty} x[n]e^{-sn} \tag{4.57}$$

if we rename our variables so that $z = e^s$, then

$$x[z] = \sum_{n=0}^{\infty} x[n]z^{-n} \tag{4.58}$$

Both the Laplace and z transforms use complex exponentials (sinusoids) as their basis functions. Although we will not cover the z-transform, it is the pathway to determine the time response of discrete systems, assess stability, design digital filters.

4.9   EXERCISES

1. Find examples of two non-science transforms. Explain what is being transformed to what. Is the transform invertible? Why or why not?

2. Transform the following differential equations to the Laplace Domain. You must generically specify the necessary initial conditions (e.g., sy(0)).

   a. $0.3\frac{d^2y}{dt^2} + 4\frac{dy}{dt} = 0.8\frac{d^3u}{dt^3} + 2\frac{d^2u}{dt^2} + u$

   b. $\frac{d^4y}{dt4} + 3\frac{dy}{dt} = 0.8\frac{d^2u}{dt^2} + 6\frac{du}{dt}$

   c. $2.5\frac{d^3y}{dt^3} - 7\frac{dy}{dt} = \frac{d^3u}{dt^3} - 3\frac{d^2u}{dt^2}$

   d. $\frac{d^2 y}{dt^2} + 2\frac{dy}{dt} + 13.25y - 5u = 0$

3. Stretch receptors in the airways of the lungs respond to stretching the wall of the airway. These receptors may be modeled using the following differential equation:

$$\frac{d^2 f}{dt^2} + 2\frac{df}{dt} + 4f(t) = \frac{dl}{dt} + 10l(t)$$

where l is the applied stretch and f is the instantaneous firing frequency of the receptor. Assume that $f(0) = -1$, $\dot{f}(0) = 2$ and $l(0) = 0$ and group f and l terms in your final answer.

   a. What is the order of this system?

   b. What is the input to this system? What is the output from this system?

   c. Write the algebraic equation for this system in the s-domain.

4. A model was derived for a circuit that detects if the heart rate has risen above 120 beats per minute. The model in the Laplace Domain is given as

$$[2s + 1]i(s) = [s^3 + 4s + 1]o(s)$$

where $i(s)$ is the input and $o(s)$ is the output in the s-domain.

   a. What is the order of this system?

   b. Write down the governing differential equation for this model.

5. Derive the following Laplace Transform for $t > 0$

$$L\left[te^{-at}\right] = \frac{1}{(s+a)^2}$$

6. One of the first models that was proposed for saccadic eye movements was proposed in 1954 by Westheimer.

$$a\frac{d^2\theta}{dt^2} + b\frac{d\theta}{dt} + c\theta(t) = f(t) \tag{4.59}$$

where θ is the eye position, $\dot{\theta}$ is eye velocity and $\ddot{\theta}$ is eye acceleration. Westheimer went on to use experimental data to fit the parameters a, b and c. Here $f(t)$ is the force applied by the extraocular muscles. Transform this differential equation into the s-domain assuming that all initial conditions are zero.

CHAPTER 5

Block Diagrams

In previous chapters we discussed mathematical models of systems. But often a system is composed of many subsystems with several signals that allow those subsystems to interact. Looking at the many differential equations (or even the algebraic equations in the s-domain) often tells you little about how the system is put together.

In this chapter we will discuss a graphical way of representing a system. *Block diagrams* are one of the cornerstones of engineering because, unlike a list of equations, you can *see* all subsystems at once and how they are connected together. You also can easily classify signals as input, output or internal.

5.1   BLOCK DIAGRAM OF A PACEMAKER-DEFIBRILATOR

Consider that the sensing system in a pacemaker-defibrilator (Figure 5.1) is composed of several subsystems. We need electrodes to collect raw signals from the heart. But these signals are small and noisy. So we need some circuitry to amplify the signal and filter out noise.

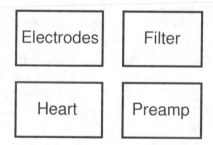

Figure 5.1: Basic components for a sensing system of a pacemaker-defibrilator.

In reality Figure 5.1 is greatly simplified. For example, there are usually stages of filtering and amplifying so there would be a few more blocks for filters and amplifiers. Also missing is the fact that often we would be recording several signals (called *channels*) from the heart at the same time that would all need to be amplified and filtered separately. If the signals are to be recorded by digital electronics, such as a computer, an Analog to Digital block would also need to be present.

You will notice that we have drawn blocks (just boxes with text in them) that represent the subsystems. You will sometimes hear engineers interchange the words subsystem and *component*. By our definition of a system, we now have a collection of *stuff*. But these isolated components don't tell the whole story and wouldn't give us anything useful (engineers would say *functionality*) if they were simply put together in the same room. To work properly they must possess the rest of the system definition and *be connected together*.

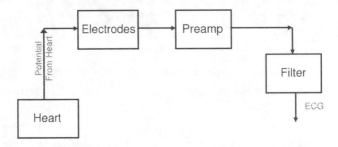

Figure 5.2: Block diagram of sensing system of a pacemaker-defibrilator showing subsystems and signals.

This redrawn figure contains blocks and arrows. The arrows represent signals that are typically a physical quantity that can be measured, in this case electrical potentials. The purpose of an arrow (signal) is to send information from one component to another in the system. In other words, they allow the various components to communicate and work together.

There are a few things to take note of in this block diagram. First, the direction of the arrow indicates the flow of information through the system. If the arrow is pointing into a block, it is an input. Likewise it if is pointing away from a block, it is an output. Again note that a single signal can be both an output from one block and an input to another block. Second, the components in a complex system often will take in one or more signals and send out one or more signals. For example, the preamp will take *in* a low amplitude signal from the electrodes and send *out* a different signal to the filter. What is not drawn in the block diagram is that there would need to be a battery (represented by a block) to power the various components. The battery has not been drawn (yet) to make the figure look cleaner, but it would be an additional input into the preamp and filter blocks.

It is now time to take a look at a more complex block diagram of the entire pacemaker-defibrillator system.

Figure 5.3 is still greatly simplified, for example not showing all the connections to and from the battery and synchronization clock, but it contains all of the major components and their connections. There are aspects of this block diagram that are worth noticing. First, there are no double-sided arrows (like you might see in a chemical reaction). It has become standard to assume that signals can only go in one direction. So for the connection between the telemetry and the microprocessor we use two one-way arrows to show that this connection is bidirectional. This is done to make it more clear how they are communicating with one another. In fact, this is a common way for two components to talk to one another. In the language of engineering we can say that both components must be able to both *send* and *receive* signals. Second, you will see that the

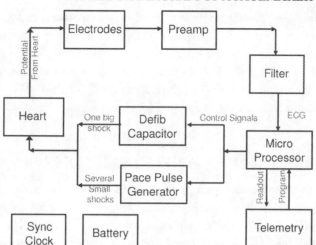

Figure 5.3: Block diagram of a pacemaker-defibrilator showing select subsystems and signals.

microprocessor may send the same control signal to both the capacitor and the pulse generator. Again, this is common in block diagrams, where two components need the same information to function, but may do different things with that information. In this case, the capacitor uses the control signal to deliver one big shock, while the pulse generator delivers a sequence of smaller pulses. Third, there is a lot going on inside of each block that we have not shown. There is a complicated circuit inside of the pulse generator and filter. There is complex computer code inside the microprocessor. And both the microprocessor and telemetry components need to have transmitter and receiver devices.

It is time to step back for a bit and reflect on what was mentioned in the introduction—Figure 5.3 graphically represents a system that would be a mess if shown as pages of differential equations, circuit diagrams and computer code. It is a very high level view of the system, engineers might say this is a *30,000 foot view*. For this reason it is often used in the design of a complex system. As an engineer, you would start with the *primary functions* for a pacemaker-defibrilator (sense an abnormality, deliver one large shock) and then start breaking out what other blocks (*supporting functions*) are needed. In fact, you could go inside of a block, for example the microprocessor, and draw another block diagram to explain all of the components inside that block. There are many different names for this type of thinking but the most common are *functional decomposition* and *modular design*.

You may have also realized that block diagrams are also an excellent way of communicating with non-engineers. It is much more intuitive than showing equations and code. It can also help in manufacturing a device to make it more clear how different components should be assembled together.

Block diagrams can be useful to show the ordering of a process as a flow chart. As engineers typically do not use signals and systems tools to analyze processes, we will not go into more depth here. The one exception might be showing how an algorithm works. So in the example above, if we were to draw a block diagram of the computer code inside the microprocessor, we would essentially be drawing a flow diagram (i.e., the flow of the algorithm). But you should realize that other fields (e.g., business, economics) use similar ideas to show the flow of goods, people, ideas or viruses.

5.2 PARALLEL, SERIES AND JUNCTIONS

There are many ways that the parts of a physical system can be connected together using blocks and arrows. Typically all of these connections are some combination of three basic connection types: *series*, *parallel* and *feedback*. We will start with series and parallel and leave feedback to Chapter 7. In all cases it will be helpful to go back to our earlier interpretation of a signal as being information that is sent either to or from a system.

A series connection is when information from one block is sent directly and unchanged to *only one other block*.

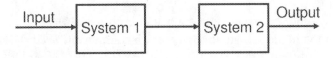

Figure 5.4: Prototypical series connection of two blocks.

A parallel connection is when information from one block is sent directly and unchanged to *more than one other block*.

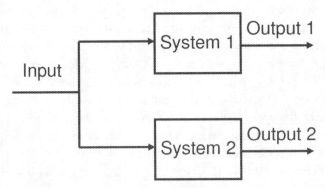

Figure 5.5: Prototypical parallel connection of two blocks.

The implications of these connections are very important when one considers the causal nature of the system inputs and outputs. For example, with a parallel connection the same information can be used in different ways. This means that several actions (outputs) may be initiated given just one input. Your body is a classic example of this type of multitasking. In your body, your heart is pumping blood, your neurons are firing, your kidneys are filtering out salts and your stomach is breaking down nutrients all at the same time. If you were out camping and encountered a bear (an input), that information would go to many systems all at once (in parallel) to raise your heart rate, send glucose to your muscles and shut down your digestive system (as well as other responses). On the other hand, if you eat a candy bar, you wouldn't want that input to turn on your entire digestive system. Instead, you would want it to only turn on your stomach. When your stomach was done, it would alert your small intestine to turn on and so on and so forth.

We know that some systems take in a lot of inputs. Others send out a lot of outputs. In Figure 5.5 we might want to combine together the two outputs into one. Most typically signals (outputs or inputs) are combined together by addition or subtraction. Rather than draw a new block to represent this simple operation, it has become standard to use a *junction*, as shown in Figure 5.6.

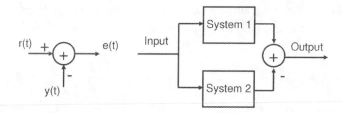

Figure 5.6: Prototypical junction on the left, with the parallel connection redrawn using a junction on the right.

Junctions perform a very simple operation—combine two or more signals to form a single output. In this case we are taking two signals, $r(t)$ and $y(t)$ and then forming a new signal, $e(t) = r(t) - y(t)$. You will notice that although the junction is given as a plus sign (addition), we can turn it into subtractions simply by putting a negative sign on the arrow coming into the junction.

You will be using series, parallel and junction connections throughout the text. Using these three connections we can build what are known as *open* systems. When we introduce feedback we will be able to create *closed* systems.

5.3 TRANSFER FUNCTIONS

So far, we have spent a lot of time with input and output signals because these are what we can measure. But how do the inner workings of the system *transform* an input into an output? In other words, what is going on inside the system? As a start we will define a *gain* as the ratio of output to input

$$Gain = \frac{out}{in} = \frac{o(t)}{i(t)} \tag{5.1}$$

where $o(t)$ is the output of our system over time and $i(t)$ is the input over time. *Gain* is what will define the system. It tells us how inputs are transformed into outputs. Throughout the text, we will use capital letters to indicate a system and lower case letters for signals. This convention can help prevent mistakes that can come in when the systems become more complex.

The idea that we will explore more in Chapter 8 is that we can put in a known input, observe the output and then form the ratio. But the idea of a gain that is a ratio of output and input is a good start and in some very simple applications will work just fine. The problem is that our signals are changing over time, so the ratio will change too. This is a problem, because we are assuming that the system is not changing even though $i(t)$ and $o(t)$ do change. We want something that will work for any system (as long it is stationary, linear and time-invariant) for all time.

The basic problem is that $i(t)$ and $o(t)$ are in the Time Domain. To move forward we will assume that our system can be described by some differential equations (either known or unknown) that govern the time behavior of the system. Let's start with the worst case where we don't know the exact differential equations. A generic linear system would be described by:

$$\frac{d^n y}{dt^n} + a_{n-1}\frac{d^{n-1}y}{dt^{n-1}} + a_{n-2}\frac{d^{n-2}y}{dt^{n-2}} + \ldots + a_1\frac{dy}{dt} + a_0 y = f(t) \tag{5.2}$$

Where n is the order of the system, the a terms are constant coefficients and $f(t)$ is called a *forcing function* or in our terms the input to the system. The output will be how our variable of interest, $y(t)$, changes over time. It is possible that $f(t)$ is also defined by a known set of differential equations:

$$f(t) = b_n\frac{d^n u}{dt^n} + b_{n-1}\frac{d^{n-1}u}{dt^{n-1}} b_{n-2}\frac{d^{n-2}u}{dt^{n-2}} + \ldots + b_1\frac{du}{dt} + b_0 u \tag{5.3}$$

Where m is the order and the b terms are coefficients. So combining equations

$$\frac{d^n y}{dt^n} + a_{n-1}\frac{d^{n-1}y}{dt^{n-1}} a_{n-2}\frac{d^{n-2}y}{dt^{n-2}} + \ldots + a_1\frac{dy}{dt} + a_0 y \tag{5.4}$$

$$= \frac{d^n u}{dt^n} + b_{n-1}\frac{d^{n-1}u}{dt^{n-1}} b_{n-2}\frac{d^{n-2}u}{dt^{n-2}} + \ldots + b_1\frac{du}{dt} + b_0 u \tag{5.5}$$

or more concisely:

$$\frac{d^n y}{dt^n} + \sum_{i=1}^{i=n-1} a_i\frac{d^i y}{dt^i} = \frac{d^m u}{dt^m} + \sum_{j=1}^{j=m-1} b_j\frac{d^j u}{dt^j} \tag{5.6}$$

In this general formulation of our unknown system, $u(t)$ is our input and $y(t)$ is our output. Here the gain as described above would be

$$G(t) = \frac{y(t)}{u(t)} \qquad (5.7)$$

This would be very difficult to compute in time. Instead we can use the Laplace Transform (see Chapter 4) to transform our differential equation into an algebraic equation. The result is:

$$s^n y(s) + a_{n-1}s^{n-1}y(s) + a_{n-2}s^{n-2}y(s) + \ldots + a_1 s^1 y(s) + a_0 y(s) = \qquad (5.8)$$
$$s^m u(s) + b_{m-1}s^{m-1}u(s) + b_{m-2}s^{m-2}u(s) + \ldots + b_1 s^1 u(s) + b_0 u(s) \qquad (5.9)$$

where we have assumed that all initial conditions are zero. This does not need to be the case and we have only done so to make the equation a bit neater. With some factoring

$$y(s)\left[s^n + a_{n-1}s^{n-1} + a_{n-2}s^{n-2} + \ldots + a_1 s^1 + a_0\right] = \qquad (5.10)$$
$$u(s)\left[s^m + b_{m-1}s^{m-1} + b_{m-2}s^{m-2} + \ldots + b_1 s^1 + b_0\right], \qquad (5.11)$$

and our definition of a Gain

$$G(s) = \frac{y(s)}{u(s)} = \frac{s^m + b_{m-1}s^{m-1} + b_{m-2}s^{m-2} + \ldots + b_1 s^1 + b_0}{s^n + a_{n-1}s^{n-1} + a_{n-2}s^{n-2} + \ldots + a_1 s^1 + a_0} \qquad (5.12)$$

We have now arrived at a useful and generic relationship, $G(s)$ is the transfer function in the s-domain. And it does not fall into the trap of changing over time. In the systems we will consider, m and n (order of the system) will be relatively small so $G(s)$ will not look nearly as intimidating.

Because the transfer functions you will usually encounter are relatively simple, they are sometimes shown on the block diagram.

Here $G_e(s)$, $G_a(s)$ and $G_f(s)$ are the transfer functions for the electrodes, preamp and filter. We could even be more specific and give the actual polynomials inside the blocks.

In Figure 5.7, the signals a-d are also in the s-domain and relate the components to one another. Given our definition of the transfer function as a ratio of output to input, we can begin to make some simple algebraic equations. By definition

$$G_e(s) = \frac{b(s)}{a(s)} \qquad (5.13)$$

We can do the same for the other blocks.

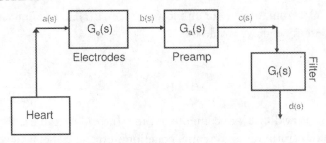

Figure 5.7: Sensor block diagram with transfer functions embedded within the blocks.

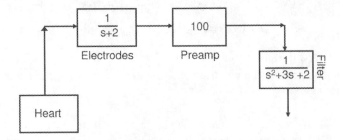

Figure 5.8: Sensor block diagram with polynomials in the s-domain embedded within the blocks.

$$G_a(s) = \frac{c(s)}{b(s)} \tag{5.14}$$

$$G_f(s) = \frac{d(s)}{c(s)} \tag{5.15}$$

Notice that all signals and transfer functions are in the Laplace (s-domain). In addition to getting away from problems in the Time Domain, we will also see that we can now deal with polynomials and simple algebraic manipulations to reduce these diagrams.

5.3.1 REDUCING BLOCK DIAGRAMS

There are times when you will only be interested in the input and output of the total system. For example in Figure 5.7, we may only care about how the signals a and d are related. In other words, what we want to find out is what the composite (or *total*) transfer function is when we pass a signal through $G_e(s)$, $G_a(s)$ and then $G_f(s)$.

And we have already done a lot of the work by writing down Equations 5.13–5.15. Because these equations are in the Laplace Domain, we can rearrange them as if they were just algebraic equations. For example, we can rewrite the first equation as

$$b(s) = a(s)G_e(s) \tag{5.16}$$

In this form, we have solved for the output, $b(s)$, in terms of the transfer function and input. We can do the same for Equations 5.14 and 5.15

$$c(s) = b(s)G_a(s) \tag{5.17}$$
$$d(s) = c(s)G_f(s) \tag{5.18}$$

and now we can recognize that what we are really after here is the total transfer function that is defined as the ratio $\frac{d(s)}{a(s)}$. But we have equations for all of these signals and all we need to do is some substitution. We can substitute Equation 5.17 in to Equation 5.18

$$d(s) = b(s)G_a(s)G_f(s) \tag{5.19}$$

and then substitute again to eliminate $b(s)$

$$d(s) = a(s)G_e(s)G_a(s)G_f(s) \tag{5.20}$$

and some rearranging will yield

$$G_{total} = \frac{d(s)}{a(s)} = G_e(s)G_a(s)G_f(s) \tag{5.21}$$

So our total transfer function is simply the multiplication (in the s-domain) of the transfer functions. And remember that in Figure 5.8 we have actual polynomials. Now we can simply combine all of these terms together.

$$G_{total} = \frac{1}{s+2} 100 \frac{1}{s^2 + 3s + 2} = \frac{100}{s^3 + 5s^2 + 8s + 4} \tag{5.22}$$

and then we can make a new reduced block diagram.

In general, the safest way to reduce a large block diagram is to write down all of the equations for the transfer functions and then use substitution. But in the sections below, we will find that there are some shortcuts that can reduce any series or parallel connection.

5.3.2   SERIES CONNECTION REDUCTION

The work for the series connection has already been done because the sensor example above is just three blocks connected together in series. When any systems are put in series, we can reduce their total transfer function down by simply *multiplying* together their individual transfer functions. This is a very powerful result and one that validates dealing with signals and systems in the s-domain.

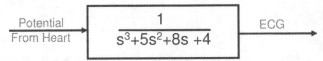

Figure 5.9: Total transfer function for the sensor system of a pacemaker-defibrilator.

5.3.3   PARALLEL CONNECTION REDUCTION

The other basic type of connection in this chapter is a parallel connection. Below in Figure 5.10 is an example of three systems that have been put together in parallel where their outputs are combined together using a junction.

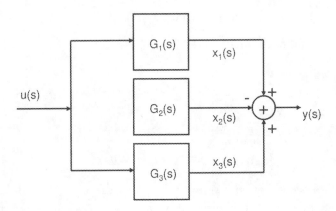

Figure 5.10: Example of three systems in parallel.

Here our goal is to find $\frac{y(s)}{u(s)}$. We can start as we did above, by writing out equations for the individual transfer functions

$$G_1(s) = \frac{x_1(s)}{u(s)} \tag{5.23}$$

$$G_2(s) = \frac{x_2(s)}{u(s)} \tag{5.24}$$

$$G_3(s) = \frac{x_3(s)}{u(s)} \tag{5.25}$$

and then rearranging to isolate the signals

$$x_1(s) = u(s)G_1(s) \qquad (5.26)$$
$$x_2(s) = u(s)G_2(s) \qquad (5.27)$$
$$x_3(s) = u(s)G_3(s) \qquad (5.28)$$

You may remember back in Section 5.2 we discussed the junction, but it was in the Time Domain (e.g., $e(t) = r(t) - y(t)$). Luckily, addition in the Time Domain is just addition in the s-domain. So we can simply combine together the signals

$$y(s) = x_1(s) - x_2(s) + x_3(s) \qquad (5.29)$$

and then we can substitute

$$y(s) = u(s)G_1(s) - u(s)G_2(s) + u(s)G_3(s) \qquad (5.30)$$
$$y(s) = u(s)[G_1(s) - G_2(s) + G_3(s)] \qquad (5.31)$$
$$\frac{y(s)}{u(s)} = G_1(s) - G_2(s) + G_3(s) \qquad (5.32)$$

Or graphically we could draw the following total block diagram

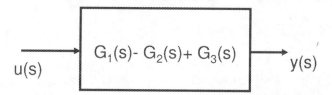

Figure 5.11: Example of three systems in parallel reduced to a single transfer function.

To abstract our result, we have found that parallel connections simply add or subtract depending upon the signs going into the junction.

5.3.4   COMBINING SERIES AND PARALLEL

Now that we know how to handle series and parallel connections, without needing to write down every intermediate signal, it is possible to revisit out original pacemaker block diagram and reduce it.

The overall input is $a(s)$ and the overall output is $z(s)$. So the overall transfer function will be

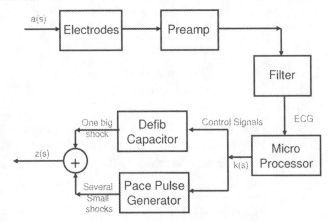

Figure 5.12: A revisit to the pacemaker block diagram but with the heart and telemetry missing and a junction added.

$$PACE(s) = \frac{z(s)}{a(s)} \tag{5.33}$$

Generally it is a good idea to do this in a few steps (i.e., not all in your head). For example, you might see that the box Electrodes is in series with Preamp, which is also in series with Filter and Microprocessor. So in one step we could combine those together to form an intermediate transfer function.

$$\frac{k(s)}{a(s)} = G_e(s)G_p(s)G_f(s)G_\mu(s) \tag{5.34}$$

Then the signal $k(s)$ is the input to two parallel systems, $G_d(s)$ and $G_{pp}(s)$. So we can write down

$$\frac{z(s)}{k(s)} = G_d(s) + G_{pp}(s) \tag{5.35}$$

Now we can just rearrange a bit

$$z(s) = k(s)\left[G_d(s) + G_{pp}(s)\right] \tag{5.36}$$

$$k(s) = a(s)\left[G_e(s)G_p(s)G_f(s)G_\mu(s)\right] \tag{5.37}$$

and then substitute

$$z(s) = a(s) \left[G_e(s)G_p(s)G_f(s)G_\mu(s)\right] \left[G_d(s) + G_{pp}(s)\right] \tag{5.38}$$

And the total transfer function is

$$PACE(s) = \frac{z(s)}{a(s)} = \left[G_e(s)G_p(s)G_f(s)G_\mu(s)\right] \left[G_d(s) + G_{pp}(s)\right] \tag{5.39}$$

Of course block diagrams can get much more complex, and you might need to use the ideas of parallel and series repeatedly, but it is much more efficient than writing down every equation.

5.4 MATLAB, SIGNALS AND SYSTEMS

Matlab has a number of commands that are useful in series and parallel connections, transfer functions, and a number of other systems concepts that will be introduced later. Nearly all of these commands are contained in the Control Systems Toolbox. In this chapter, we will introduce the most basic commands and then expand upon them in later chapters.

The most basic command is the *tf* (for Transfer Function). The idea is that if you have the function derived above

$$G_{total} = \frac{1}{s+2} 100 \frac{1}{s^2 + 3s + 2} = \frac{100}{s^3 + 5s^2 + 8s + 4} \tag{5.40}$$

can be represented in Matlab with the following commands

```
>> Num = [100]; % Create Vector for Numerator
>> Den = [1 5 8 4]; % Create Vector for Denominator
>> G = tf(Num,Den) % Create the transfer function
```

Here we have created two vectors that contain the coefficients for the numerator and denominator of the transfer function. You should note that you sometimes need to use zeros to get the coefficients right (e.g., $s^3 + 4s$ would be represented as the vector [1 0 4 0]). The last command creates the transfer function. We have purposely left off the last semicolon so that Matlab will print out the transfer function to the command line.

The transfer function above was derived from three blocks in Figure 5.8 that were put in series. We could have used Matlab to do the work for us, using the *series* command.

```
>> Electrodes = tf([1],[1 2])
>> Preamp = tf([100])
>> Filter = tf([1],[1 3 2])
>> Temp1 = series(Electrodes,Preamp);
>> TotalG = series(Temp1,Filter)
```

A few things should be noted here. First, in creating a transfer function, you can simply put the two vectors (Numerator and Denominator) into the call to *tf* (e.g., tf([1],[1 2])). Second, when a transfer function only has a numerator (as in Preamp), we only specify one vector. Third, we build up the total transfer function using the *series* command in steps—we define Temp1 first and then use it in the final step.

Matlab also has a command, *parallel*, that will combine together systems in parallel. For example, we could reduce the diagram in Figure 5.10. Let's assume that

$$G_1(s) = \frac{1}{s+3} \tag{5.41}$$

$$G_2(s) = \frac{s}{s^2 - 4} \tag{5.42}$$

$$G_3(s) = \frac{200}{10s + 3} \tag{5.43}$$

First, we can define the three transfer functions

```
>> G1 = tf([1],[1 3])
>> G2 = tf([1 0],[1 0 -4])
>> G3 = tf([200],[10 3])
>> Temp1 = parallel(G1,-G2)
>> TotalG = parallel(Temp1,G3)
```

Note that to change the sign of G2 we simply put a "-" sign in the fourth command. Hopefully, you will agree that the mess that comes out of this was much easier to compute than if you did it by hand!

5.5   EXERCISES

1. Draw the dynamics of learning in your class using blocks and arrows. You should include yourself, your instructor, all of your assignments, projects and labs as well as any feedback loops.

2. Draw a block diagram of a non-science system. What might be measurable in this diagram? How would you measure it? Clearly identify the inputs and outputs.

3. Given the following biological signals:

   a. Hormone in the blood stream

   b. Electrical impulse in the brain

   c. Mechanical contraction of a muscle cell

   define three different systems such that the signal is

i. an input

ii. an output

iii. an internal signal within a larger system (please do not say "the body").

Show your nine systems as block diagrams.

4. Give the total transfer function of Figures 5.13 and 5.14. For Figure 5.13 you will be able to write down the transfer function by eye using the shortcuts shown in this chapter. For Figure 5.14, you will need to go back to the original way of finding transfer functions (i.e., writing down systems of equations). Note that there is a new *feedback* connection (arrows that point backward toward the input). We will revisit this idea more in the Chapter 7. For both block diagrams, you should express your answer as a transfer function, $\frac{b(s)}{a(s)}$ that contains $F(s)$, $G(s)$ and $H(s)$.

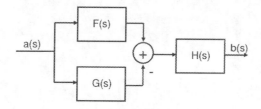

Figure 5.13: Block diagram for a series and parallel system.

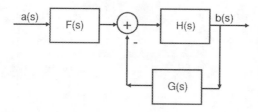

Figure 5.14: Block diagram for a feedback system.

5. Assume that the sub-systems in problem 4 have the following transfer functions

$$F(s) = 35$$
$$G(s) = \frac{s}{s-4}$$
$$H(s) = \frac{1}{s+2}$$

Find the total transfer function for both block diagrams. You are encouraged to use Matlab to do your calculations. Whether you use Matlab or not, please show your work (calculations or Matlab commands).

CHAPTER 6

Stability

6.1   INTRODUCTION

In previous chapters we learned that we can think about systems in either the time or the s-domain, and that each has advantages and disadvantages. In this chapter, we develop a way to determine if a system is stable or not.

The word "stability" has a meaning in everyday English. But as engineers, we mean something very precise by stability. To us, there are different ways to be stable, different ways to be unstable and even a spectrum between stability and instability. We can also use our methodologies to predict what will happen as we add parts to a system.

Before diving into the math, it is important to understand that the stability of a system is about a preferred or *natural* behavior. For example, we can think of a bridge as being stable because all the forces balance in such a way that the entire structure is *statically stable*. That means that it is not changing over time—all the forces balance in the same way now and in the foreseeable future. Systems may also be *dynamically stable*—while they do change over time they have some preferred output. The common theme in both types of stability is that the output of these systems never blows up to infinity. Any system that has the potential to have an infinite output will be called unstable.

So what do we mean by natural behavior? Here the phrase Bounded Input, Bounded Output (or *BIBO* for short), is helpful. What BIBO means is that given a bounded input, the system will produce a bounded output. The definition of stability is stated this way because we can always make a system unstable by sending in an infinite (or at least very large) input. For example, if we put an infinite voltage across an electrical resistor it will most certainly be unstable. But for reasonable inputs (and the meaning of reasonable depends on the system) a stable system will not blow up.

But this still doesn't quite explain what we mean by "natural" behavior. What we really mean is that if a system is moved away from its natural behavior (an engineer would call this a *perturbation*) the system will find its way back (an engineer would call this a *transient*) to its natural behavior. So the way to test for stability in a dynamic system is to see if any bounded perturbation will result in a transient that eventually leads back to a characteristic natural behavior.

6.2   STABILITY AND TRANSFER FUNCTION POLES

In this section we will explore how properties of the transfer function in the s-domain can be used to measure system stability. The key will be to find what are known as the *poles* of the system. So before we get into stability, we will first explain what a pole is and how to find one from a transfer function.

6.2.1   FINDING POLES AND ZEROS

Remember that the transfer function is defined as the ratio of input to output in the s-domain. This typically results in a function with a numerator and denominator that have some number of s terms. Recall the equation from Chapter 5

$$G(s) = \frac{s^m + b_{m-1}s^{m-1} + b_{m-2}s^{m-2} + \ldots + b_1 s^1 + b_0}{s^n + a_{n-1}s^{n-1} + a_{n-2}s^{n-2} + \ldots + a_1 s^1 + a_0} \tag{6.1}$$

We can interpret the numerator and denominator as polynomials of the variable s. And we can factor the polynomials and rewrite $G(s)$ as:

$$G(s) = \frac{(s - z_1)(s - z_2)\ldots(s - z_m)}{(s - p_1)(s - p_2)\ldots(s - p_n)} \tag{6.2}$$

Given this factorization we can easily find the roots of the numerator

$$z_1, z_2, \ldots, z_m \tag{6.3}$$

and denominator

$$p_1, p_2, \ldots, p_n \tag{6.4}$$

In system terminology, the roots of the denominator are called the *poles* while the roots of the numerator are called the *zeros*. Engineers call the value of n (i.e., number of poles) the *order* of the system, regardless of the value of m. So if a system has one pole (e.g., a simple RC circuit), it is a first-order system. If it has two poles (e.g., a simple RLC circuit), it is second-order.

For any system with an order greater than one (i.e., transfer function denominator has s^n terms where $n > 1$), there is the possibility of having complex poles. In other words a pair of solutions that looks like

$$p_{n-1,n} = A \pm Bj \tag{6.5}$$

where A and B are constants and j is imaginary. For more information on complex numbers, see Appendix A. We will return to the meaning of complex poles later in the chapter. The same idea can also be applied to the zeros, but we will not worry about complex zeros.

6.2.2   VISUALIZING POLES AND ZEROS

Rather than simply list the poles and zeros, engineers, being the graphical people that we are, will often create a plot with pairs of A's and B's. In other words we will plot the poles and zeros on the Real-Imaginary axis as mentioned in Chapter 4 (also called the complex plane).

Let us assume that we have two poles at -1 ± 3, a *double pole* at -3 and a zero at 3. To plot the complex poles, we place an "X" at the point (-1,3) and (-1,-3), as in Figure 6.1. The convention for a pole is always an "X." To plot the real double pole we place an "X" at (-3,0) with a "2" above the symbol to designate that it is a double pole. Now this double pole has no imaginary part so it falls right on the real axis. Likewise if we had a pole that only had an imaginary part, it would fall on the imaginary axis. To plot the zero at 3, we place a "0" on the real axis at (3,0). The convention for zeros is a "0."

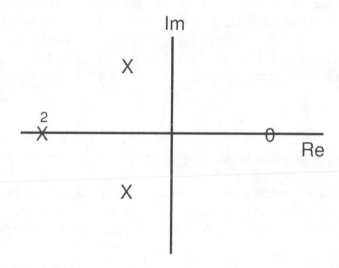

Figure 6.1: Visualization of poles at -1 ± 3, a double pole at -3, and a zero at 3.

And you should be able to see that the polynomial in the s-domain and the graphical representation are interchangeable. So if someone gave you only Figure 6.1, you could substitute the value of the poles and zeros into Equation 6.2

$$G(s) = \frac{s - 3}{(s - (-1 + 3j))(s - (-1 - 3j))(s + 3)(s + 3)} \tag{6.6}$$

$$G(s) = \frac{s - 3}{(s + 1 - 3j)(s + 1 + 3j)(s + 3)(s + 3)} \tag{6.7}$$

6.2.3   RELATIONSHIP TO STABILITY IN TIME

It is now time to explore more deeply why poles govern the stability of a system. The most straight-forward is to recall Equation 4.53 where the time solution looks something like

$$l(t) = a_1 e^{\lambda_1 t} + a_2 e^{\lambda_2 t} + \ldots + a_n e^{\lambda_n t} \tag{6.8}$$

The λ exponents are simply the roots of what is called the *characteristic equation* and in a math class called an *eigenvalue*. In fact the characteristic equation is just the denominator of our transfer function! And the λ terms are exactly the same as our poles!

Using Equation 6.8 we can figure out how our poles influence stability. For example, if all of our poles (λ's) are negative real numbers, then the exponentials will decay. In other words, every term will decrease over time and eventually the output will head toward some stable value.

What is more is that the negative value of the λ's will control how fast the time solution will decay. Large negative values of λ will decay quickly while small negative values will decay more slowly. This rate of decay is what an engineer would call a *time constant*. A time constant is simply the inverse of a pole because it governs the rate of decay of a perturbation.

$$\tau_1 = \frac{1}{p_1} \tag{6.9}$$

It will be helpful to remember that the rate of decay is inversely proportional to how far a pole is away from the origin in Figure 6.1.

We can go one step further. For a system with multiple poles, and therefore multiple λ's, in Equation 6.8 we can say a bit more about the time solution. Imagine that you have a system with one very large negative pole (far from the origin, λ_f) and one small negative pole (close to the origin, λ_s).

$$l(t) = a_1 e^{-\lambda_s t} + a_2 e^{-\lambda_f t} \tag{6.10}$$

A bit of thinking will show that the first term is going to decay *slowly*, while the second term will have a *fast* decay. So we have two decay rates (time constants), but overall the system will largely obey the fastest decay, or the largest λ.

But now consider what happens if even one of the poles (λ's) is positive, say λ_p. The exponential with λ_s will decay to some value and so will the exponential with λ_f. But the term $a_3 e^{\lambda_p t}$ will eventually grow to infinity over time because λ_p is positive. What we can conclude is that all it takes is one pole to be positive and the system will be unstable. What is more, we can even get some idea of how fast a system will blow up. Here again, the largest positive eigenvalue will dominate how quickly the blowup occurs.

If we look at Figure 6.1 we don't see any positive poles (remember that zeros don't matter for stability), so we can conclude that this system is stable. But we know that our example system does have complex poles. And other systems might even have purely imaginary poles (an "X" right on the IM axis). What do we do with these types of systems?

First we will consider what happens in time if a pole lands right on the imaginary axis (say at $2j$). And remember that because imaginary (or complex) poles must always occur in pairs, we really have two imaginary poles ($\pm 2j$). Substituting into Equation 6.8 we will have a term that looks like

$$a_1 e^{2jt} + a_2 e^{-2jt} \qquad (6.11)$$

But we know from previous chapters (and Appendix A) that when we see an exponential raised to some imaginary power, we have a sinusoid of some sort. So the conclusion we can reach is that if a pair of poles is purely imaginary it will cause the output to oscillate! This behavior is not stable in the same way as a decaying solution—it doesn't blow up but doesn't settle down either. So we give it the special name: *marginally stable*. The reason for this name is that if the system changes a little (say one coefficient in Equation 6.1) it might move the poles left (now the system is stable) or right (now the system is unstable). So you can think of marginally stable systems as the exact point where a system crosses over from stable to unstable.

The relationship between imaginary λ's and the sinusoid takes on a special meaning in systems that oscillate. Consider the following equality

$$\cos(\theta t) = \frac{e^{\theta jt} + e^{-\theta jt}}{2} \qquad (6.12)$$

What this tells us is that imaginary poles take on the role of θ, and therefore control the rate (or frequency) of the oscillation. The larger the imaginary pair of poles (farther from the origin) the faster the oscillations. The smaller the pole pair (closer to the origin) the slower the oscillations.

We still haven't addressed what happens if the pole is a complex number composed of both real and imaginary parts, as it is in our example system. Here we will have more complicated behavior in time. Let's see what happens if we simply plug a complex pair of poles into Equation 6.8. For the pair -1 ± 3

$$a_1 e^{(-1+3j)t} + a_2 e^{(-1-3j)t} \qquad (6.13)$$

Because of how exponentials work we can split out the real and imaginary parts

$$a_1 e^{-t} e^{+3jt} + a_2 e^{-t} e^{-3jt} \qquad (6.14)$$

Each term has a decaying portion and an oscillating portion. Now consider that the oscillation will stay bounded, but the decay will head to some small value. The two will fight it out, but in the end the decay will win, making the amplitude of the oscillations smaller and smaller. What we will get is a sort of *ringing*. Just like a bell, when a system with a complex pair of poles is "hit" (an engineer would say perturbed) it will ring for a while. But eventually the oscillations will die out.

Lastly, we can consider what would have happened if our poles were 1 ± 3. In other words, if the real part was positive. If you plug these poles in you will find that now, instead of a decaying exponential and an oscillator, you have an increasing exponential and an oscillator. Using similar logic, this system will ring but in reverse—it will oscillate small at first, but with the oscillations growing larger and larger over time. This system will be unstable.

At this point you should be able to look back at Figure 6.1 and say that it is stable (although its transients might oscillate a bit).

We are now ready to come to the big conclusion of this chapter. A system is stable if *all* of its poles are on the left side of the imaginary axis (an engineer would say *in the left half plane*). A system is unstable if *only one* pole is in the right half plane. We can summarize what we have found graphically.

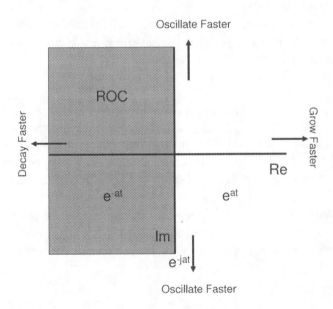

Figure 6.2: Complex plane showing the relationship between poles, stability and rates of decay, growth and oscillation.

You will notice the shaded region which has been labeled ROC for *Region of Convergence*. This is to indicate that any system that has all poles in the left half of the complex plane will be stable and the natural solution will eventually *converge* (either to a number or a bounded oscillation).

6.3   THE ROLE OF ZEROS

While the poles are important in determining the system stability, the zeros are important too. Above we thought of a time solution as being a sum of exponentials. But we left the door open to weighing those exponentials by arbitrary constants (a's). So it may be that some exponentials count more than others because their amplitudes are bigger. What the zeros do, at least in part, is help set these constants.

Consider that the numerator of the transfer function, $G(s)$, is a polynomial. There is a method called *partial fraction expansion*, introduced in Chapter 4 and explained in more detail in Appendix B, where the transfer function can be rearranged into the form

$$G(s) = \frac{c_1}{(s - p_1)} + \frac{c_2}{(s - p_2)} + \ldots + \frac{c_n}{(s - p_n)} \tag{6.15}$$

Here we can see that numerators (zero) impact the relative magnitudes of growth or decay (poles).

6.4   DESIGNING SYSTEMS

As engineers, we are nearly always concerned with making a stable system. In fact, in the typical case, the more stable the better. When trying to design a system it is therefore important to have some idea of where the poles are. Better yet, we could think about design as placing poles (and zeros too) in the appropriate place.

On the other hand, there are some systems which should oscillate (think of the wiper blades on your car, or your heart beat). By placing poles on the imaginary axis, you can design a system that oscillates. And, by designing where they fall on the imaginary axis, you can even control the oscillation frequency. There is a problem here. If anything changes inside the system (and things always do in a biological system), you could easily wind up with an unstable system.

It may come as a surprise that many very functional systems are in fact unstable! A person on two legs is often modeled as an inverted pendulum (a skinny rod with a weight at the top)—a very unstable configuration. Why would you design a system this way? And how can it function if it is unstable? The answer is very instructive in thinking about designing systems. We will only give a partial answer in this chapter, but will expand on it in chapters to come.

Rather than think about the transient (decay, growth, oscillations), let's consider what happens once a system reaches a stable point. By definition it will stay there. And it will try to resist perturbations, by decaying back to the stable point. This sounds pretty boring and inflexible. In some sense the system will not be able to explore very much, and when it does, it will only be motivated to change by some outside perturbations. Biological systems and biomedical devices need to be less restrictive. They need to be able to expand outward from a natural stable point at times. And an unstable system is just the answer. It allows for a falling away from stable points, just as walking is falling away from a static standing posture.

But we wouldn't want to continue to be unstable forever. What we want to find is some trick for allowing a naturally unstable system to start blowing up but then to catch itself before it actually does. This is in the same spirit as thinking of walking as controlled falling. The falling part is unstable, but the catching part (when you put your foot down) prevents the overall system from becoming entirely unstable (you falling over). The next chapter on feedback will discuss how this is possible.

6.5   MATLAB AND STABILITY

Matlab has a few built-in functions that can be helpful in assessing the stability of systems with complicated transfer functions. First, if we know the poles, we can create a polynomial. The following Matlab commands

```
>> Num = poly([1+j,1-j,-1]);
>> Den = poly([1+2j,1-2j,-2]);
>> F = tf(Num,Den);
```

create a transfer function with zeros at $1 \pm j$ and -1, and poles at $1 \pm 2j$ and -2. The *poly* command here is used to create a polynomial given the roots.

$$F(s) = \frac{s^3 - s^2 + 2}{s^3 + s + 10} \tag{6.16}$$

The reverse function in Matlab is *roots*—given the roots it will find the coefficients of a polynomial. If we had been given $F(s)$ we could find the poles and zeros by

```
>> Zeros = roots([1 -1 0 2])
>> Poles = roots([1 0 1 10])
```

And we can plot the poles and zeros of a transfer function with

```
>> pzmap(F)
```

And we can get the values of the poles and zeros by

```
>> [Poles, Zeros] = pzmap(F)
```

6.6   EXERCISES

1. For the differential equations below

   a. $0.3\frac{d^2 y}{dt^2} + 4\frac{dy}{dt} = 0.8\frac{d^3 u}{dt^3} + 2\frac{d^2 u}{dt^2} + u$

   b. $\frac{d^4 y}{dt^4} + 2\frac{d^3 y}{dt^3} + 3\frac{dy}{dt} = 0.8\frac{d^2 u}{dt^2} + 6\frac{du}{dt}$

c. $\frac{d^4y}{dt^4} + 2.5\frac{d^3y}{dt^3} - 7\frac{dy}{dt} - 4y = \frac{d^3u}{dt^3} - 1.541\frac{d^2u}{dt^2}$

d. $\frac{d^4y}{dt^4} + 2.5\frac{d^3y}{dt^3} - 7\frac{dy}{dt} - 4y = \frac{d^3u}{dt^3} - 3\frac{d^2u}{dt^2}$

e. $\frac{d^2y}{dt^2} + 2\frac{dy}{dt} + 13.25y = 2\frac{du}{dt} - 5u$

i. Write down the transfer function in the s-domain $\frac{y(s)}{u(s)}$, assuming zero initial conditions.

ii. Report the zeros and poles. You may use Matlab.

iii. Report if the system is stable, unstable or marginally stable.

2. For the following transfer functions

a. $\frac{6s-5}{14s^2+7s+1}$

b. $\frac{s^3-2s+3}{8s^2-2}$

c. $\frac{s}{s^4+5s+10}$

i. Write down the governing differential equations.

ii. Draw the location of the zeros and poles n the s-plane. You may use Matlab.

3. Given the following poles, compare the time response (qualitatively) to a system with poles at -1 and $-1 \pm j$. You should compare any oscillations, decay or growth. Please explain your answers.

a. $-3, -5$

b. $-1, -1 \pm -2j$

c. $\pm 3j$

d. $-4, 2, \pm 3j$

CHAPTER 7

Feedback

You will very often see or hear the term *feedback* when dealing with systems. In fact, all but the simplest systems have some sort of feedback. This is especially true in biological systems and devices that interface with the body. As this chapter progresses, you should understand why most every system contains feedback. Very briefly, there are two primary reasons. First, feedback can be used to allow an inherently unstable system to remain stable. Second, feedback opens the door to controlling the output of a system. We will explore stability in this chapter and control in the next.

7.1   OPEN AND CLOSED LOOP SYSTEMS

Imagine that you are listening to your favorite song on a nice stereo system. Consider the block diagram of your hearing

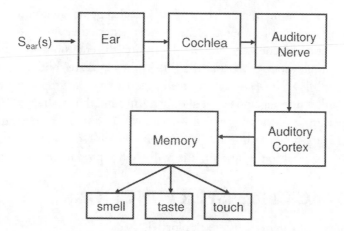

Figure 7.1: Block diagram of processing of auditory input.

From Figure 7.1, we might speculate as to why this signal is your favorite song. You will notice that we can follow the signal from the ear to the cochlea to the auditory nerve, then to the auditory cortex, all as serial connections. From here the signal is passed along to your memory where it is recognized and associated (sent as parallel connections) to other sensory areas of the brain. These various connections can be strengthened and are the reason that this grouping of sound waves is your favorite song.

Using the techniques we learned in Chapter 5 we could label all of the subsystems and signals and then reduce the block diagram down to a single transfer function, $G_{hear}(s)$. You should notice that the signals all go in one direction from the ear to memory. A system where all the signals flow in one direction is known as an *Open Loop* system and $G_{hear}(s)$ would be called the *Open Loop Gain* of the system.

In reality, your sense of hearing is much more complex. Consider that you may be at a crowded party and your favorite song is being played. You can easily tune in to that sound even if other things in the room are louder. Likewise, you may want to focus your attention on a conversation you are having and ignore the annoying song that comes on next. What we need is some way for the gains inside of $G_{hear}(s)$ to dynamically change. If your brain were like a mixing board, their would be a series of dials and knobs to make these dynamic changes. But in the case of your brain the knobs are actually inside the system! At some fundamental level, that is what feedback does. It allows a system to pay attention to what is happening and then adjust accordingly.

Before we get too carried away with our analogy, we need to recognize that simple feedback is just that—simple. For example, consider what a thermostat does. It measures the temperature of the room. If it is too hot, it will send a signal to the air conditioner to turn on. As cool air is coming into the room, the temperature reading goes down. When it reaches a desirable temperature, the signal to the air conditioner is cut off. Here the feedback is created because a thermostat has a way to measure its environment, but also has a way to change its environment.

The thermostat example is one that involves an external environment. We could say that this is an *external* kind of feedback—one that loops into the environment and then back into the system. The ear example seems a bit different—all of the loops are contained within the brain. So we could say that this is an *internal* form of feedback. But we can stop and pause for a minute to reflect on the idea that internal and external only mean something when we draw a box around a system. If in the example of the thermostat we defined the system as the entire room plus the air conditioner, then everything was internal. So while it is helpful sometimes to think of the feedback loop as internal or external to the system, it is really only a product of the system definition.

7.2   FEEDBACK TRANSFER FUNCTIONS

Let's now take a look at a generic feedback block diagram.

Now there are two inputs to the system, some input, $u(s)$, that is external to the system (i.e., from outside the system boundary) *and* the current output, $y(s)$. That output is sent back through another internal system, $H(s)$, and that signal is subtracted from the input. $G(s)$ is often called the *Open System* while $H(s)$ is called the *Feedback System*. Notice that now we have some signals that move from left to right and at least one other signal that moves right to left. Again, this feedback may be very obvious and exposed (e.g., the thermostat in a heating and cooling system), or it may be buried deep inside of a system (e.g., the way your body controls your level of attention).

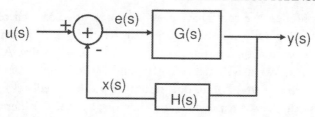

Figure 7.2: Block diagram of a basic feedback system.

If all we care about is the open part of the system then we can simply cut the feedback connection to find the *open loop gain*. This open loop gain is simply the transfer function found by following the signal from left to right, starting with the input, $u(s)$, and ending with the output, $y(s)$. In this case it will just be $G(s)$.

But what is the transfer function when we put the feedback back in? What we need to find is $\frac{y(s)}{y(s)}$, also known as the *closed loop gain*. We already derived general equations for serial and parallel connections and now we will do the same for a generic feedback connection.

We may start by writing down the three equations that relate our signals to our systems.

$$y(s) = e(s)G(s) \tag{7.1}$$
$$e(s) = u(s) - x(s) \tag{7.2}$$
$$x(s) = y(s)H(s) \tag{7.3}$$

Next we must solve these equations for $\frac{y(s)}{u(s)}$:

$$y(s) = [u(s) - x(s)]G(s) \tag{7.4}$$
$$y(s) = [u(s) - (y(s)H(s))]G(s) \tag{7.5}$$
$$y(s) = G(s)u(s) - y(s)H(s)G(s) \tag{7.6}$$
$$y(s)[1 + H(s)G(s)] = G(s)u(s) \tag{7.7}$$

and finally

$$\frac{y(s)}{u(s)} = \frac{G(s)}{1 + H(s)G(s)} \tag{7.8}$$

Equation 7.8 is known as Black's Formula and may be used for any type of feedback block diagram.

You may have wondered why we subtracted $x(s)$ going into the junction. It is because this was a generic diagram for *negative* feedback. You will see a bit more about what negative feedback does below, but typically it will take $u(s)$ and minimize its impact on the system (literally subtract out some of its amplitude). We could have given $x(s)$ a positive contribution to the junction in which case the block diagram would be *positive* feedback. In general positive feedback will tend to increase the impact of $u(s)$ on a system.

This idea is exactly what we need to understand at a conceptual level what is needed for your auditory system to pay attention to one input (use positive feedback), while reducing other inputs (negative feedback). As a teaser of future chapters, there is still something else that needs to happen—if all of the sound is coming in through the ears in one signal, how do you amplify some parts but not others? You will get more clues as you read more.

We can now take a look at a concrete example

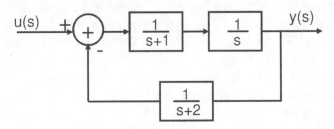

Figure 7.3: Block diagram of a basic feedback system.

The system in Figure 7.3 is not exactly like the generic feedback system in Figure 7.2, so we will need to do a little work first. There really is only one difference—instead of a single open loop system, we have two. But these two systems are just in series and we know how to reduce systems in series. After collapsing the $1/(s + 1)$ and $1/s$ systems we will have

$$G(s) = \frac{1}{s(s + 1)} \tag{7.9}$$

$$H(s) = \frac{1}{s + 2} \tag{7.10}$$

and we can apply Black's Formula:

$$\frac{y(s)}{u(s)} = \frac{\frac{1}{s(s+1)}}{1 + \frac{1}{s(s+2)(s+1)}} \tag{7.11}$$

$$\frac{y(s)}{u(s)} = \frac{s + 2}{s(s + 2)(s + 1) + 1} \tag{7.12}$$

So the open loop gain is simply $\frac{1}{s(s+1)}$ but the closed loop gain is:

$$\frac{y(s)}{u(s)} = \frac{s + 2}{s^3 + 3s^2 + 2s + 1} \tag{7.13}$$

7.3 BLOCK DIAGRAM REDUCTIONS

The power of Black's Formula is that it may be used in combination with the reduction techniques for series and parallel from Chapter 5. Consider the following block diagram

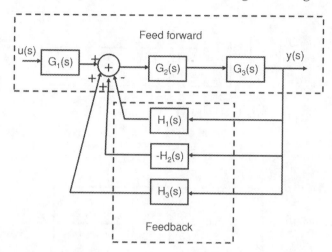

Figure 7.4: Block diagram with series, parallel and feedback connections.

We could certainly write down all of the signals and then algebraically solve through the equations for the transfer function $\frac{y(s)}{u(s)}$. But it is much easier to reduce systems using what we know about series, parallel and feedback connections. It is also good to notice that there is an open loop system and a feedback system. The open loop system is made up of the subsystems that form the open loop. And they are all in series. So the open loop gain of this system is simply $G_1(s)G_2(s)G_3(s)$.

To start, we can reduce the two systems that lie between the output and the junction.

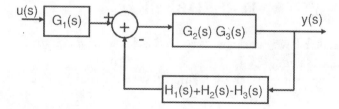

Figure 7.5: Block diagram with one series reduction and one parallel reduction.

In Figure 7.5 there are two systems in the forward path, $G_2(s)G_3(s)$. It is important to note that we can't yet reduce $G_1(s)$ because the junction is in the way. The feedback systems are all in parallel (they all start and end in the same place). This reduction (taking into account signs) is $H_1(s) + H_2(s) - H_3(s)$. Here we have thought ahead a bit. Because this system has feedback, we know we will use Black's Formula at some point. And because Black's Equation is for negative feedback, it will make things simpler in the end if we flip around the various signs of $H(s)$ to make it negative feedback.

We now can use Black's Formula to reduce everything except for $G_1(s)$

u(s)
G1(s)
G2(s) G3(s)
1+ G2(s) G3(s) [H1(s)+H2(s)-H3(s)]
y(s)

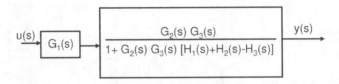

Figure 7.6: Block diagram reduced using Black's Formula.

Finally we can combine our two series systems.

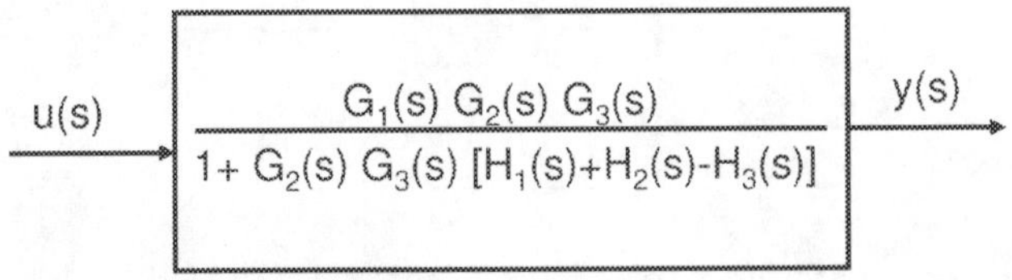

u(s)
G1(s) G2(s) G3(s)
1+ G2(s) G3(s) [H1(s)+H2(s)-H3(s)]
y(s)

Figure 7.7: Reduced transfer function for Figure 7.4.

In general you will want to start with the obvious parallel, series and feedback subgroups and work your way outward. There are times when it may not be obvious what to do next. In these cases, you may need to go back to the algebraic way of finding a transfer function.

7.4 STABILITY AND FEEDBACK

At the end of Chapter 6 we mentioned that feedback was a way to turn an inherently unstable system, into a stable system.

Consider an open loop system in Figure 7.8 with the transfer function

$$G(s) = \frac{1}{s-2} \tag{7.14}$$

The pole of G(s) is at 2 and therefore the system is unstable. For the sake of argument, let's assume we have a simple feedback system, as in Figure 7.9 with a feedback transfer function of $H(s) = K$, where K is some constant.

Using Black's Formula

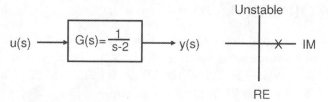

Figure 7.8: An inherently unstable system.

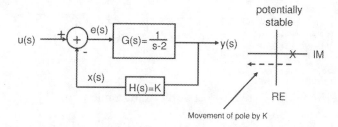

Figure 7.9: An inherently unstable system made stable by feedback.

$$\frac{y(s)}{u(s)} = \frac{\frac{1}{s-2}}{1 + \frac{K}{s-2}} \tag{7.15}$$

$$\frac{y(s)}{u(s)} = \frac{1}{s - 2 + K} \tag{7.16}$$

Consider what happens if $K = 1$. Here our transfer function becomes

$$\frac{y(s)}{u(s)} = \frac{1}{s - 1} \tag{7.17}$$

and the system is still unstable with a pole at 1. But if we continue to increase K, say to a value of 3

$$\frac{y(s)}{u(s)} = \frac{1}{s + 1} \tag{7.18}$$

the system is now stable with a pole at -1. So changing the constant K is changing how much the feedback matters. In the s-domain we see that it is pushing the pole to the left (more stable).

This is a very rudimentary form of feedback, similar to a thermostat. What we really want is some more dynamic way to adjust the system as we go. It would be as if the system itself could detect what is needed to remain stable. This will be the topic we will take up in Chapter 9.

7.5   FEEDFORWARD

In feedback, we moved a signal in the direction from output back to input. But there is another way to move a signal forward that is called *feedforward*. In the most basic type of feedforward, we leapfrog a signal over some number of subsystems, as in Figure 7.10. But why do this? There could be many reasons, and we will discuss just one here.

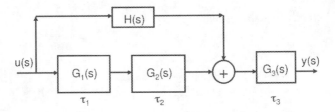

Figure 7.10: An example of a system with a feedforward connection.

Imagine that you have a system, $G_3(s)$, that has built into it some delays. We can think of these delays simply as a time constant, τ_3. The meaning is that this system doesn't react right away when it starts getting an input—it takes some time to get it going. Now imagine that we have two other systems $G_1(s)$ and $G_2(s)$ each with their own time constants (delays). If we simply moved through these three functions, we would need to wait a long time ($\tau_1 + \tau_2 + \tau_3$) for a significant output, $y(s)$, to appear.

Now, if we add the subsystem $H(s)$ in feedforward, and it has a short delay, then while G_1 and G_2 are working on the signal, $H(s)$ can be *priming* G_3 to receive a signal. It is analogous to priming a pump in case you might need it quickly. So feedforward can get a subsystem ready for signals that will be coming.

In biological systems we can even go a step further. Imagine that $H(s)$ is actually a filter of some sort—it will pass only parts of $u(s)$ (or maybe nothing at all). In this case, it could help to amplify (positive feedforward) or inhibit (negative feedforward) the input to $G_3(s)$. This is in fact how parts of your vision pathways work. When a new visual stimulus is presented, your brain will do some processing, but it is also waking up and priming other areas of the brain that might be needed later on in the processing.

7.6   OPENING THE LOOP

As mentioned in the introduction, nearly all biological systems have feedback integrated into them at all levels. This poses a problem for anyone (usually a scientist) trying to figure out what is going on. The reason is that feedback has a way of messing up causality. Consider that in the generation of an action potential in a neuron or muscle we have a membrane voltage that is arrived at as a difference in potential between the inside and outside of a cell membrane. If ions should move across this membrane, the membrane voltage would change. And in fact there is a way for these ions to get across the membrane—through ion channels that are made up of proteins. These proteins can change the flow of ions (current) that cross the membrane, and in doing so, they can control the membrane potential. But here is the catch: the level to which the channels are open is governed by the membrane voltage. So we have this loop where membrane voltage controls current, but the current controls the membrane voltage.

And, from this example, it is therefore impossible to say that one is causing the other. For that reason, science has developed some tricks. We will talk about one in this chapter and then go into more depth in the next chapter. The trick for this chapter is to recognize that we can restore causality by simply cutting this bidirectional feedback. In effect, allowing voltage to impact current, but not the other way around. That in fact is exactly what Hodgkin and Huxley did in a clever experiment that won them the Nobel Prize in 1962. They were able to fix the membrane voltage, but then allow the current to change. In essence, they took a closed system and turned it into an open system.

7.7   MATLAB AND FEEDBACK

The Matlab controls toolbox has a command for feedback. The command is called *feedback* and can handle positive and negative feedback. To demonstrate, we can revisit Figure 7.3. First we can define the three sub-systems.

```
>> Sys1 = tf([1],[1 1]);
>> Sys2 = tf([1],[1 0]);
>> FeedbackSys = tf([1],[1 2]);
```

Then we can find the open loop gain (the two systems in series)

```
>> OpenLoop = series(Sys1,Sys2);
```

Then we can use the *feedback* command to get the total transfer function

```
>> CloseLoop = feedback(OpenLoop,FeedbackSys)
```

Note that the feedback command automatically assumes negative feedback. If we were to change this system to positive feedback, the command would be

```
>> PositiveFeedback = feedback(OpenLoop,FeedbackSys,+1);
```

7.8   EXERCISES

1. List two examples of feedback in biological systems. Explain whether they are examples of positive or negative feedback.

2. List two examples of feedback in a social, economic or political system. Explain whether they are examples of positive or negative feedback.

3. Derive Black's Formula change for positive feedback.

4. Derive a formula for the feedforward diagram in Figure 7.10.

5. For the three systems shown in Figures 7.11–7.13

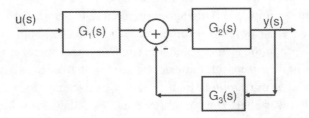

Figure 7.11: Feedback homework problem 1.

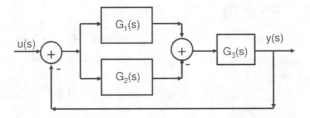

Figure 7.12: Feedback homework problem 2.

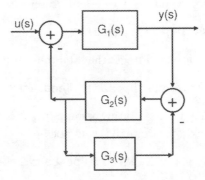

Figure 7.13: Feedback homework problem 3.

Assume that

$$G_1(s) = s + 2 \tag{7.19}$$

$$G_2(s) = \frac{1}{s-3} \tag{7.20}$$

$$G_3(s) = \frac{s+1}{s^2} \tag{7.21}$$

For each system

a. What is the open loop gain? Show both a generic answer, using G_1, G_2 and G_3, as well as an analytic answer with polynomials of s. You are encouraged to use Matlab.

b. Is open loop system stable?

c. What is the closed loop gain? Show both a generic answer, using G_1, G_2 and G_3, as well as an analytic answer with polynomials of s. You are encouraged to use Matlab.

d. Is the closed loop system stable?

6. Pacemaker cells in the heart initiate a rhythmic wave of electrical activity that propagates throughout the heart. In the most simplistic view, these cells have developed a way to become marginally stable and oscillate at a particular frequency.

a. Design an open loop system (transfer function) that will oscillate at 1Hz. It is helpful to know that the imaginary part of a pole is in units of radians per second (rad/sec).

b. Draw a closed loop feedback loop around your system where $H(s)$ is equal to a constant K. Find the transfer function and explain why K cannot be used to create a stable system

c. Design $H(s)$ such that the heart rate will become 2Hz.

CHAPTER 8

System Response

We have established that a dynamic system can be described using a mathematical model that can be represented as a transfer function in the s-domain. But if we are given a transfer function, what outputs will it produce? This is known as a *forward problem* in physics and math because we are given a generic system and want to find how the system behaves. In engineering, this is like designing a system that can perform certain functions. But what can we do if we are given a system but don't know anything about it (i.e., we don't have any equations or a transfer function)? In physics and math, this is known as an *inverse problem*. In some sense what we are trying to do is reverse engineer a system. Typically we think of scientists as being concerned with the inverse problem (figuring out the system) and engineers as being concerned with the forward problem (given the equations figuring out how the system will behave).

In reality, engineers should be able to do both forward engineering and reverse engineering. In both forward and reverse engineering we need a set of tools to, on the one hand, demonstrate and quantify the behavior of a known system, and on the other hand characterize an unknown system. There are a number of ways we already know to help us characterize a system. For example, in Chapter 6 we learned how poles determine the order and stability of a system. But we assumed that the system had reached an equilibrium. In this chapter we want to know about the *transients* of the system—how it behaves when it is not in equilibrium. The general method by which this is done can be summed very simply—send a known signal into the system and observe what comes out.

An analogy may help. If you enter a dark room with a flashlight, your first instinct is to send your light beam to different parts of the room. Here you are *mapping* out the space by sending a signal in (narrow beam of light, as an input) and measuring the result (response of your eyes and processing by your brain, as the outputs). Even though you might never see the room all at once, after a bit of time, you could probably draw a very accurate picture of what was in the room.

Now there is a catch to our analogy. Most rooms are very rich and non-linear places—each part looks different. All of the systems we have encountered so far will be linear and time-invariant. So if we send in several inputs at once, our output will simply be the sum of the output from each individual input. In this chapter we will send in a few very general inputs that will allow us to learn a great deal about the system. It will be like shining the flashlight into the room once and then being able to make a map of the rest of the room. These generic inputs are the *impulse*, *step* and *sinusoid* that we explored in Chapter 4.

8.1   ZERO INPUT AND ZERO STATE RESPONSE

The Zero Input Response (ZIR) is how a system behaves when there is no input, $x(s) = 0$. Although this may seem unusual there are three reasons to consider the ZIR. First, $y(s) = H(s)x(s)$ and it would seem that since $x(s) = 0$ there should be no output. Consider, however, that there are domains of the system where $H(s) \to \infty$. These are exactly at a pole. Second, the system may oscillate if a pair of poles are purely imaginary. This oscillation will happen in the absence of any input.

The third, and perhaps most important use of the ZIR is to determine how a system decays from some given initial condition. Consider the circuit in Figure 8.1. If we place our initial condition at $V_o = 5V$ and let the system evolve, we will see that the time response will decay (exponentially) to 0V. We will learn more about this circuit in later sections.

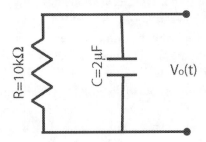

Figure 8.1: A simple RC circuit that might be an analog of a mechanical, thermal or other system.

The Zero State Response (ZSR) is the response of the system to *any* input when the initial conditions are set to zero. In practice, this is often difficult to achieve in a real situation because most complex biological systems or devices are not in equilibrium and so the dependent variables are never set to zero. The reason for the definition of ZSR is because the total response of a system can be defined as:

$$Y_T(s) = ZIR + ZSR \tag{8.1}$$

In reality, the zero input and zero state responses are theoretically interesting, but they are hard to use in practice. They are, however, terms that are useful to know.

8.2   THE IMPULSE RESPONSE

The impulse response of a system is found by waiting until the system has reached a stable equilibrium and then giving it a very precise nudge. We can then observe how the output of the system finds its way back to equilibrium. Recall that the impulse is defined as in Figure 8.2.

$$\delta(t) = \begin{cases} \infty & \text{if } |t| \leq \frac{t}{2} \\ 0 & \text{Otherwise} \end{cases}$$

By definition the area under this infinitesimally short burst is 1.

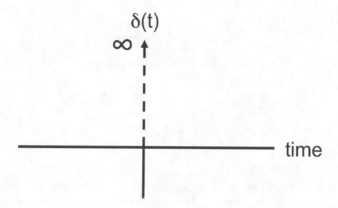

Figure 8.2: Impulse function, $\delta(t)$, used to evaluate a system.

The reason for sending in an impulse is that the Laplace Transform is very simple

$$L[\delta(t)] = 1 \tag{8.2}$$

Let's consider what happens when we assume that our input to a system is $x(t) = \delta(t)$. Remember that the transfer function is defined as:

$$H(s) = \frac{y(s)}{x(s)} \tag{8.3}$$

$$H(s)x(s) = y(s) \tag{8.4}$$

But for an impulse $x(s) = 1$ so

$$H(s) = y(s) \tag{8.5}$$

or the system output *is* the system transfer function in the s-domain. This is a very profound result, because it means that with one simple input, we can get the transfer function of a linear system. In the Time Domain we call this the *impulse response*, or $h(t)$.

8.2.1   A FIRST ORDER EXAMPLE

Imagine that we have a very simple first-order RC circuit, as shown in Figure 8.1. In the s-domain

$$H_L(s) = \frac{1}{RCs + 1} \tag{8.6}$$

$$H_L(s) = \frac{1}{0.02s + 1} \tag{8.7}$$

The time constant of the system ($\tau = 0.02$) is just the value of RC. And so there is a pole at -50. The Time Domain solution therefore is

$$V(t) = V_o e^{-50t} \tag{8.8}$$

where V_o is the initial condition.

What this tells us is that if we put a voltage on this circuit (say $V_o = 25V$) that it will decay exponentially over time from V_o. Now an impulse is a bit different because we are assuming that the circuit starts with $V_o = 0$, but then we apply a short but very intense voltage pulse. The result can be found in Figure 8.3. Note that the impulse jumps the system to $V = 50$. This is because a large impulse is really like setting an initial condition.

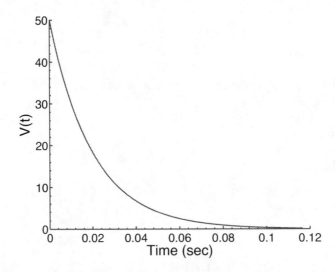

Figure 8.3: The impulse response of an RC circuit with a time constant of 0.02sec.

This is a classic impulse response, $h(t)$, from a stable first-order system. It is also a demonstration of a forward problem—we know the transfer function and we want to find an output.

Let us assume that an impulse is delivered to this system, but the transfer function is unknown. In other words, all you have is Figure 8.3. From this curve you might guess that it is just a single exponential decay so this is a first-order system with only one time constant.

$$V(t) = V_o e^{-t/\tau} \tag{8.9}$$

$$V(t) = V_o e^{-t/RC} \tag{8.10}$$

But we know one more bit of information because we know $V(t = 0) = V_o = 50$ from the graph. So at $t = 0$

$$V(0) = V_o e^0 = 50 \tag{8.11}$$

$$V(t) = 50 e^{-t/\tau} \tag{8.12}$$

Now we need to find the time constant, τ. Here we can do a little bit of a trick. Let's assume that we are at $t = \tau$

$$V(t = \tau) = 50 e^{-\tau/\tau} \tag{8.13}$$

So at $t = \tau$ the exponent goes to -1

$$V(t = \tau) = 50 e^{-1} \tag{8.14}$$

$$V(t = \tau) = 50 e^{-1} = 50 \times 0.37 = 18.4 \tag{8.15}$$

What this means is that V=18.4 when $t = \tau$. We can estimate when that happens in Figure 8.3 and sure enough, it happens for $t = 0.02$! This is an example of an inverse problem—working from $h(t)$ to a system equation.

8.2.2   A DIFFERENT FIRST ORDER EXAMPLE

The derivation above was a first-order RC circuit that acts as a lowpass filter (more will be covered on filters in Chapter 12). Note that in Figure 8.1 the resistor and capacitor are in parallel. If we put them in series, the circuit will be a highpass filter. A typical transfer function (with the same R and C values) will look like

$$H_h(s) = \frac{RCs}{RCs + 1} \tag{8.16}$$

$$H_h(s) = \frac{0.02s}{0.02s + 1} \tag{8.17}$$

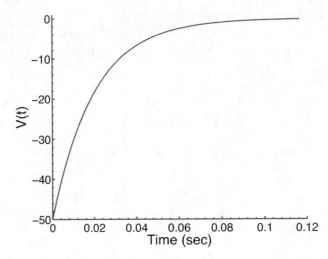

Figure 8.4: The impulse response of an RC circuit with a time constant of 0.02sec.

We can again apply an impulse and observe the result, shown in Figure 8.4. This is still an exponential, but the Time Domain solution is a bit different

$$V(t) = V_o e^{-t/\tau} \tag{8.18}$$
$$V(t) = -50e^{-t/\tau} \tag{8.19}$$

where we have used the same trick of setting $t = 0$ to find V_o. And we can then use the same trick of setting $t = \tau$ to get the time constant.

$$V(t = \tau) = -50e^{-\tau/\tau} \tag{8.20}$$
$$V(t = \tau) = -50e^{-1} \tag{8.21}$$
$$V(t = \tau) = -50e^{-1} = 50 \times 0.37 = -18.4 \tag{8.22}$$

Again, by setting $t = \tau$ we can read the time constant off of the impulse response, in this case when $V = -18.4$ at 0.02 seconds.

8.2.3   A SECOND ORDER EXAMPLE

As we saw in previous chapters, a more complex result can occur if we have a second (or higher) -order system. Let's assume our system has the transfer function

$$H(s) = \frac{1}{s^2 + 0.5s + 1} \tag{8.23}$$

By factoring the denominator we can find that this system has poles at $-0.25 \pm 0.9682j$. From Chapter 6 we can say that this system will be stable because the real part is in the left half plane. And we know that because there is an imaginary part, the system will have some oscillations. Figure 8.5 shows what happens when we start the system at equilibrium and then apply an impulse.

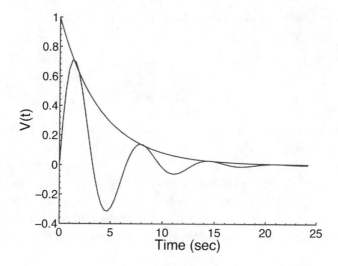

Figure 8.5: The impulse response (black) for a second-order system, along with the envelope (red) for the decaying exponential.

The impulse response is plotted in black in Figure 8.5 and it does have decaying oscillations. But we can go even further, given that we know the poles. Plotted in red is the *envelope* of the decaying exponential

$$Env = e^{-0.25t} \tag{8.24}$$

In other words, the real part of the pole will give you the envelope of the decay.

And as you may have guessed, because the imaginary part of the pole is what causes the oscillation, it is that value that will tell you the frequency of oscillation. But the imaginary part, ± 0.9682, is in radians/second. To get this in terms of our usual frequency measure (1 Hertz = 1 cycle / second) we need a conversion.

$$\omega = 2\pi f \tag{8.25}$$

$$f = \frac{\omega}{2\pi} \tag{8.26}$$

where ω is in radians/second and f is in Hertz. So for our system, the oscillations should be at a frequency

$$f = \frac{0.9682}{2\pi} \tag{8.27}$$

$$f = 0.1541\,Hz \tag{8.28}$$

What can help is to know that the period (T) is the inverse of frequency

$$T = \frac{1}{f} \tag{8.29}$$

$$T = \frac{1}{0.1541} = 6.5s \tag{8.30}$$

So the cycle should repeat every 6.5 seconds. If you take a look at the peaks (or valleys) in Figure 8.5, you can see that they do in fact occur every 6.5 seconds.

As in the first-order examples, this was the forward approach—we knew the transfer function and wanted to find the response. But we can also take an inverse approach—what if we were given nothing but Figure 8.5? You should be able to guess that it is stable (because it does not blow up) and that it has a complex pole (because it has a decaying oscillation) and you could even estimate the real and imaginary parts of the poles from the decay rate and period of oscillation.

8.3   THE STEP RESPONSE

The impulse response perturbs a system from equilibrium but then allows the system to return to that equilibrium. The *step response* does not allow the system to return to equilibrium, but forces the system to stay away from its natural equilibrium. The simplest way to achieve this is to apply some very long lasting and constant perturbation. In a stable system this will usually result in a new force equilibrium. Observing this new equilibrium state and how the system gets there (transient) may provide more information about the underlying system. In theory, a long time would be infinity but we are engineers so we need a different definition of "a long time." Practically, we continue the perturbation until the new equilibrium is reached.

In theory we defined a step function in Chapter 4 and shown in Figure 8.6.

$$u(t) = \begin{cases} 1 & \text{if } t > 0 \\ 0 & \text{Otherwise} \end{cases}$$

You should note that sometimes the step response is also called the *Unit Step* or *Heaviside*. But as an engineer, we often might apply something other than a unit step. For example, we might apply 5V to a circuit rather than 1V. But that doesn't matter as long as we know that our system is linear because the general behavior will simply scale.

8.3.1   THE IMPORTANCE OF THE STEP RESPONSE

In Chapter 4, we derived the Laplace Transform of a step response as

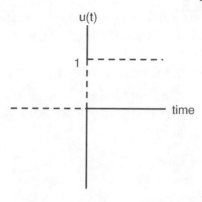

Figure 8.6: Step function, $u(t)$, is zero before $t = 0$ and one after $t = 0$.

$$L[u(t)] = \frac{1}{s} \qquad (8.31)$$

We can go through the same analysis as the impulse function where $x(t) = u(t)$.

$$H(s) = \frac{y(s)}{x(s)} \qquad (8.32)$$

$$H(s) = \frac{y(s)}{1/s} \qquad (8.33)$$

$$y(s) = \frac{H(s)}{s} \qquad (8.34)$$

So the response is not quite as simple as the impulse but it is still fairly simple. Why not simply always use an impulse function to characterize a system? There are two reasons, one practical and one more theoretical. Practically, it is much easier to deliver a good representation of step functions then an impulse. For an impulse we need to deliver a pulse that is really short. But the meaning of really short in this case is going to be relative to the time constants of the system. And if we don't know the time constants ahead of time, then we don't know how short to make the impulse. With a step we need to keep the perturbation constant for a long time. Again this is relative, but at least now we know that we have left it on long enough when the system settles to a new steady-state.

The more theoretical idea is that some systems are of high-order, meaning that they might have some slow and fast time constants. Physically, you can think of these as systems with elements that can store some quantity of the system (e.g., capacitors and inductors, reservoirs, chemical buffers). And some will impact the system in the short term and others in the long term. An impulse often will not expose these differences in time constant. But the step response can. Again, this is not precise, but there are cases where the step can give you more information.

8.3.2   COMPARING THE STEP AND IMPULSE RESPONSES

A natural question would be to find out how the impulse and step response differ from one another. To begin to answer this question the step response for the first-order high filter described in Equation 8.17 is shown in Figure 8.7.

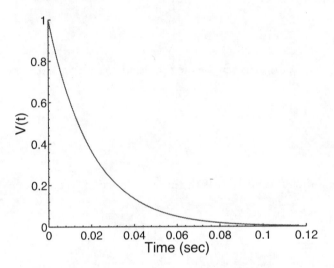

Figure 8.7: The step response for the transfer function in Equation 8.17.

There is clearly a qualitative difference between Figures 8.3 and 8.7. But the key point here is that we can read the same time constant off of both plots ($\tau = 0.02s$).

Why would they be different? One way to think about this is that all the impulse does is set an initial condition. So this really is like solving a homogeneous differential equation. The step function on the other hand is an initial condition but also includes a constant forcing function. In other words, the time solution is going to be from both a homogeneous and a particular solution.

We can now do the same sort of analysis for the second-order step response in Figure 8.8. With the step response of the second-order transfer function we get what we might expect—a decaying oscillation—but with one difference. We start with an initial value of zero, then rise to some peak and oscillate around the value of 1.

8.4   QUANTIFYING A RESPONSE

All of these differences in possible responses (and we haven't even seen the full range yet) lead to the question of how to generically characterize the response of the system. While we cannot

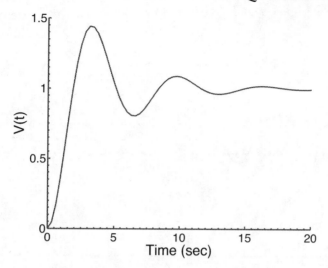

Figure 8.8: The step response for the transfer function in Equation 8.23.

anticipate every possible measurement, there are some standards that are used among engineers that are good to know. Typically they are based off of a second-order system that has decaying oscillations, like in Figure 8.8. But that response was from a somewhat special transfer function. Below in Figure 8.9 is the response from an unknown transfer function and the definition of some relative measures.

Steady-State is the value that the system reaches at $time \to \infty$. A bit of thought will show you that the system behavior is tied to the poles of the transfer function. Just like the poles, there are three possibilities: 1) All poles are negative and the steady-state is a finite value. The steady-state value will then tell us something about our system. 2) Some poles are purely imaginary and the response oscillates. In this case there is no steady-state. Here the rate of oscillations will tell us something about where the poles are on the imaginary axis. 3) At least one pole has a positive real part and the impulse response goes to infinity. In Figure 8.9 we can say that the system is stable because it does settle down to the value of 6.

Overshoot is the maximum positive deflection from the steady-state when the steady-state value is a finite number. This nearly always occurs right after the impulse. For example, if the steady-state is 6 and the maximum positive deflection is 10, then the overshoot would be 4. Note that we have reported the overshoot *relative* to the steady-state. There are in fact, two ways to report the overshoot. In our example, it may be reported as an absolute number, 4, or as a percentage of the steady-state, 66.6% ($100\left[\frac{10-6}{6}\right]$).

Undershoot is the maximum negative deflection from the steady-state when the steady-state value is a finite number. The measurement is similar to overshoot. For example, if the steady-

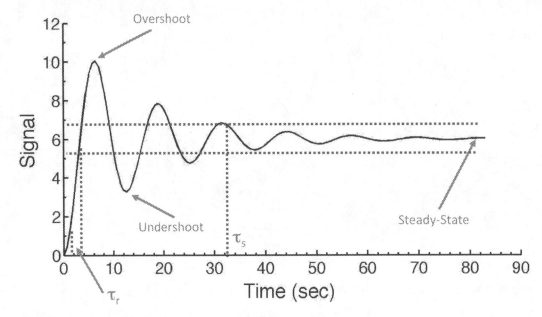

Figure 8.9: The step response for an unknown transfer function.

state is 6 and the maximum negative deflection is 3, then the undershoot would be reported as either 3 or 50%.

It is important to be careful with undershoots and overshoots. They should always be measured *after* the signal has made an initial crossing of the steady-state. So, in our example, we wouldn't count 0 as the undershoot.

Rise Time (τ_r) is a measure of how quickly the response moves from an initial condition to the steady-state. Although the method of measuring rise time may vary, a typical measure is from 10% of the steady-state to 90% of the steady state. So we are looking for the time required for the system to move from 0.6 to 5.4. In our system we might estimate this to be something on the order of 2 seconds. The rise time is a good measure of how quickly the system can respond to changes in an input (e.g., a step input). In a mechanical system this might tell you something about the inertia of the system. In an electrical system rise time is an indication of the time to charge and discharge various storage elements in the system.

Settling Time (τ_s) is a measure of how quickly any oscillations (overshoots and undershoots) die out. This is done by measuring the time that is required for the response to *always* stay within some bounds (horizontal lines in Figure 8.9). Again there are different methods of measurement, but a typical measure would be when the oscillations are some percentage of the steady state. Often the percentages used are 2, 5 or 10%. In our example we might use 10% and get a time of a little over 32 seconds.

Note that it is possible to have some systems where some of these measurements cannot be made. For example, if there are no overshoots or undershoots then we cannot define the settling time.

8.4.1   ESTIMATING A TRANSFER FUNCTION

Consider now the inverse problem of having Figure 8.9 and then trying to write down a transfer function. First we need to assume that our system is second-order and therefore has two poles. And we see that there are oscillations so the poles will be a complex conjugate pair. We can start with the oscillations which will give us the imaginary part. To do this it is easiest to estimate the time between two peaks. If we pick the first and second overshoot we have times of about 6 and about 20. This means the period of the oscillation is 14 seconds. But now we need to turn this into a frequency in radians/sec. First we can get it in Hertz (1/sec) by taking the inverse of the period. That gives us a frequency of 0.0714Hz. The last step is to convert to radians/sec using Equation 8.25.

$$\omega = 2\pi(0.0714) = 0.4486\text{rad/sec} \tag{8.35}$$

The next step is to find the real part of the transfer function. This is controlled by the envelope that is formed by the decaying oscillations. To estimate this we can use a few tricks. First we can try to draw the envelope and then try to find a function that will fit it.

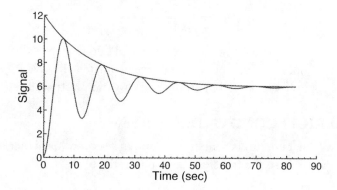

Figure 8.10: The step response for an unknown transfer function.

Here we can think of a function of the form

$$V(t) = 6 + 6e^{P_1 t} \tag{8.36}$$

This function was simply eye-balled knowing the initial value and the final steady-state value. So the next step is to pick a point off of the graph and then find P_1. It is often easiest to pick the first overshoot (or first undershoot). In this case, we can estimate that $V(t = 8) = 10$.

$$V(t = 8) = 6 + 6e^{P_1 8} = 10 \tag{8.37}$$

and with some rearranging and use of the natural log.

$$4 = 6e^{P_1 8} \tag{8.38}$$

$$\frac{4}{6} = e^{P_1 8} \tag{8.39}$$

$$ln\left[\frac{4}{6}\right] = 8P_1 \tag{8.40}$$

$$-\frac{1}{8}ln\left[\frac{4}{6}\right] = P_1 \tag{8.41}$$

$$P_1 = -0.0507 \text{seconds} \tag{8.42}$$

So we would estimate that the poles are -0.0507 ± 0.4486.

It is now time to reveal the actual transfer function that was used to generate the step response in Figure 8.10.

$$G(s) = \frac{s + 6}{4s^2 + 0.5s + 1} \tag{8.43}$$

If you factor the denominator and find the poles you will find that they are at -0.0625 ± 0.4961. So we are somewhat off, but given how roughly we estimated the values (we could have done much better if we could have zoomed in on the plot), we did pretty well.

8.4.2   A GENERIC SECOND ORDER SYSTEM

There is another way that engineers sometimes think of a second (or higher) -order system. That is to rearrange some terms so the transfer function is in the form of

$$G(s) = \frac{\omega_n^2}{s^2 + 2\delta\omega_n s + \omega_n^2} \tag{8.44}$$

Now on the surface this seems to be more complicated. But the terms ω_n and δ now have a physical meaning; ω_n is the *natural frequency* and δ is known as the *damping factor*. To understand why, we can think about what the poles of this system will be. So we need to use the quadratic equation you learned in middle school to the denominator to find the roots r_1 and r_2

$$G(s) = \frac{\omega_n^2}{(s - r_1)(s - r_2)} \tag{8.45}$$

$$r_{1,2} = \frac{-2\delta\omega_n}{2} \pm \frac{\sqrt{4\delta^2\omega^2 - 4\omega_n^2}}{2} \tag{8.46}$$

$$r_{1,2} = -\delta\omega_n \pm \omega_n\sqrt{\delta^2 - 1} \tag{8.47}$$

We can start by considering $\delta = 0$ which is known as an *undamped* system. When this is the case, the poles are $\pm\omega_n j$ which means that we have a system with purely imaginary roots. By our previous analysis this means the system will naturally oscillate. And now you should be able to see that ω_n controls the rate at which the undamped system will oscillate, and why it is called the natural frequency.

We can also consider when $\delta > 1$. When that is the case the square root will be a positive value and therefore the roots will have some purely real value. This means that the behavior of the system will have no oscillations and will follow some exponential decay like in Figure 8.3. Engineers call this an *overdamped* system because the damping is strong enough to dampen any natural oscillations.

Next we can consider when $\delta < 1$. In this case the square root will be negative and the roots will have some imaginary component. So the system will have decaying oscillations like in Figure 8.5. Engineers would call this an *underdamped* system because the damping is light and allows some of the natural oscillations to occur.

The dividing line between overdamped and underdamped occurs when $\delta = 1$. This is called a *critically damped* system. In reality you won't encounter a system that is exactly critically damped (except maybe in theory). But some specification sheets for products will give information on critical damping as a simple way to show where this dividing line occurs.

8.5 THE SINE RESPONSE

We saw how powerful the impulse and step responses could be to quantify the response of a system or transfer function. But the impulse response looks at a single very fast perturbation, and the step is a sustained input. Another worthwhile bit of information is to know how the system behaves when the input is some time-varying signal. The sine response is intended to determine how a system responds to inputs that are changing over time.

We saw in Chapter 4 that a sinusoid can be written in complex notation as

$$x(t) = Ae^{j\omega t} \tag{8.48}$$

If we assume that our system is linear and time-invariant, an interesting phenomenon occurs when we input a sinewave. A little thought will reveal that the output must also be a sinewave with the *same frequency*. Even if the system has some delay, *everything* will be delayed by the same amount. If you are not convinced, you can certainly crank through the math where the system will determine the homogeneous solution and the sinusoid input will determine the particular solution. Therefore, if a sinewave is put into the system, only the amplitude and phase of the output will change.

But now we run into a problem. Unlike the impulse or step response, there are an infinite number of sinewaves, all with different frequencies, amplitudes and phase shifts that we could send into our system. The amplitudes are not a problem because this is a linear system and the phase shifts aren't either because we are assuming our system is time-invariant. But we still need to deal with the possibility of an infinite number of frequencies to test.

Obviously we don't want to create an infinite number of time plots. But knowing that we don't need to actually show each frequency will help us create a plot that will contain all the amplitude and phase information we need. But, before we do, we need to cover one other topic that will be used to create this plot.

8.5.1   DECIBELS

Although the phase of the output sinewave will only vary between 0 and 360 degrees, the output amplitude could vary over an enormous range. Our bodies encounter a similar problem. Most notably, the intensity of sound and light can vary over a large range. Our senses have come up with an ingenious way of compressing a large range into a small range—our ears and eyes use the idea of a logarithm. We will use the same idea to compress the range of a signal. To do so, we will define *decibels* (dB) as

$$dB = 10 \log(A^2) \tag{8.49}$$

Where A is the amplitude of the sinewave. According to a law of logarithms:

$$dB = 20 \log(A) \tag{8.50}$$

To see how logarithms compress a range consider the table below:

A	dB
.001	-60
.01	-40
0.1	-20
1	0
10	20
100	40
1000	60

We have compressed the amplitude over a range of one million to a range of 120. Remember that an amplitude of 1 results in dB of 0.

8.5.2   THE BODE PLOT

We can now come back to the idea of representing the output of a system to a sinusoidal input with any frequency, all in one plot. We already established that the frequency won't change, but the amplitude and phase may. So to condense all of our time plots we will plot the input frequency versus the output amplitude and phase, as in Figure 8.11.

The *Bode* plot is actually two plots, one for amplitude and one for phase. Typically we can make this plot by sending in a sinewave with an amplitude of 1 and a phase shift of 0. Doing so makes it easy to interpret the output. The interpretation of the top plot is that the y-axis will tell us the gain (output amplitude/input amplitude) at a specific frequency (x-axis). The interpretation of the bottom plot is that the y-axis will tell us the phase shift at a specific frequency.

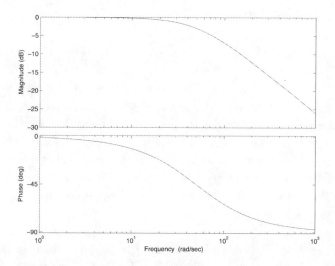

Figure 8.11: Bode plot for the RC circuit shown in Figure 8.1.

Figure 8.11 is the frequency response for the RC circuit (Figure 8.1) we analyzed earlier. First you will notice that the bottom axis shows the frequency in radians/sec of the input sinewave. But it is in decibels. In other words, the range of 0 to 10 takes up the same space on the x-axis as 10 to 100. Also, the bottom axis is often in powers of ten, known as a *decade*. The tick marks between decades are not uniform, but follow a pattern. Let's consider the interval between 10^1 and 10^2. The first tick mark past 10^1 has the value of 2×10^1 or 20 radians/second. The next tick mark is 3×10^1 or 30 radians/second, and so on until we get to 10^2. The other decades follow the same pattern.

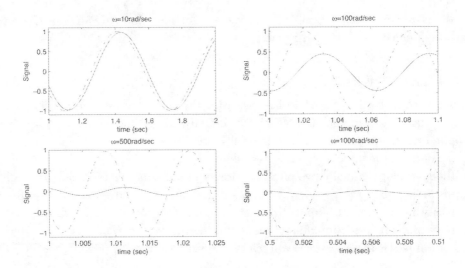

Figure 8.12: Bode plot for the RC circuit shown in Figure 8.1.

We will now connect the Bode plot to the actual output responses. In Figure 8.12 the panels show the response (in blue) to inputs (in red) at ω = 10, 100, 500 and 1000 radians/second with an amplitude of 1 and phase shift of 0. The theme of these plots is that the output amplitude decreases as we turn up the frequency. For example, at ω = 10rad/sec the amplitude of the input and output are nearly the same. There is a very slight phase shift, but it is basically zero. Now look back at the Bode plot in Figure 8.11. If you find the input frequency of 10^1 and then look at the magnitude you will see that it is zero decibels. But remember that the way we define decibels (as a logarithmic ratio of output to input) means the gain is 1—in other words the magnitude of input and output should be the same. We can do the same for the phase plot and we see that the shift is not quite zero but it is close.

If we continue to look at the Bode plot we see that as the frequency increases, the magnitude of the output should go down. And looking at Figure 8.12 we see that this is the case. The same goes for the phase. So the Bode plot contains all of the information that would be needed to draw the output of *any* sinewave input.

Consider a sinusoidal input to the circuit in Figure 8.1 that has an amplitude of 5V, frequency of $10Hz$ and phase shift of 30 degrees. To use the Bode plot in Figure 8.12 we first need to convert the frequency from Hz to rad/sec using Equation 8.25. This calculation means that we need to read off values of phase and magnitude at 62 rad/sec. At this frequency the Bode plot shows a decrease in magnitude of -5dB and a phase shift of -45 degrees. These changes, however, are relative to the input signals. So the output will be -5dB $X 5V$ or 2.81V. The phase shift is -45 degrees but that is relative to the shift already in the input. So the output will be shifted -75 degrees. That means that the output will be at 10Hz with an amplitude of 2.81V and phase shift of -75 degrees.

8.5.3   THE -3DB POINT

From the Bode plot we can derive an important point, known as the -3dB *point*. Here we can get a preview of some later material by substituting $s = j\omega$ into Equation 8.6. The idea that we will explore in more detail later is that the s in the s-domain can be interpreted as frequency. To get at the amplitude we can find the real part of this substitution (i.e., take the absolute value).

$$\left| G(s) \right| = \left| \frac{1}{1 + j\omega RC} \right| = \frac{1}{\sqrt{1^2 + (\omega RC)^2}} \tag{8.51}$$

Now we can ask what happens if $\omega = RC$, or connecting to the previous section, what happens when $\omega = \frac{1}{\tau}$?

$$\left| G(s) \right| = \frac{1}{\sqrt{1^2 + 1}} \tag{8.52}$$

$$\left| G(s) \right| = \frac{1}{\sqrt{2}} = 0.707 \tag{8.53}$$

This is also sometimes called the *half power* frequency. You might wonder why it is only "half" power. The reason is that power is calculated as a squared term and $(0.707)^2 = 0.5$.

Remember that on a Bode plot the magnitude is usually in dB so

$$20 \log \left[\frac{0.707}{1} \right] = -3\text{dB} \tag{8.54}$$

What this means is that the frequency at which the magnitude drops to the -3dB point *is* the value of RC! Let's take a closer look at where this occurs in Figure 8.11. If you find the -3dB point on the magnitude plot and then read down to find the frequency you will find that it is about 50 rad/sec. So $RC = 50$ and that means the time constant is $\frac{1}{RC} = 0.02sec$. This shows a very deep relationship between the pole, time constant and -3dB frequency for a first-order system.

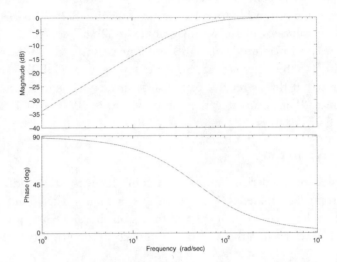

Figure 8.13: Bode plot for the highpass RC circuit.

Figure 8.13 is the Bode plot for the highpass RC circuit. We can again find the -3dB point, and from that the time constant and location of the pole of this first-order system. But, as you may have guessed from the name, these circuits are called high and lowpass for a reason. In the lowpass, we noticed that as the frequency was turned up the circuit would decrease the amplitude, in effect killing the amplitude of the output. The highpass does just the reverse—letting through high frequencies with a gain of 1, but killing the amplitude of low frequencies. We will use this idea in Chapter 12 to build filters.

So far, we have seen that the magnitude either rises or falls as we turn up the frequency. But, the second-order system we studied above (Equation 8.23) has a surprise.

The second-order Bode plot in Figure 8.14 starts out by passing low frequencies (gain is 0dB), but then rises! This means that the output amplitude is *greater* than the input amplitude. We will not get into exactly how this happens (although you should be suspicious that the laws of conservation of energy are being violated; they are not). What is striking is that around 1 rad/sec there is a peak, after which increasing the frequency causes the output amplitude to drop. We will save further analysis of Bode plots like this one for Chapter 12, but there is one more term that is worth defining and remembering.

The peak in output amplitude that occurs for some range of frequencies is known as the *resonant* frequency. Some second-order systems resonate (rings louder) when certain frequencies are sent in. A classic example of resonance is riding over a speed bump. You may have discovered

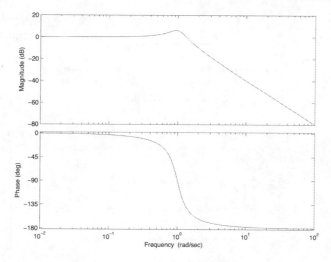

Figure 8.14: Bode plot for the second-order system in Equation 8.23.

that if you ride over a speed bump very slowly, the shocks on your car have time to absorb the shock and you do not get jerked around after the bump (your car oscillating up and down). Likewise, if you ride over it quickly, you sort of float over the bump and again your car does not oscillate up and down. But there is a speed (often "normal" driving speed) for which the speed bump has a big impact on your car. This is because a speed bump sends in a resonant frequency to your car as the front and then back wheels go over the bump.

8.6   RESPONSE TO AN ARBITRARY INPUT

We have taken a look at the impulse, step and sinusoidal inputs, but in reality most systems take in some arbitrary input from their environment. The power of assuming a linear and time-invariant system is that once we know the response to one impulse, we can send in any combination of impulses and compute the output. As a thought experiment, imagine sending in a number of impulses, as in Figure 8.15. Put another way, we can add impulses together or scale them in any way we want and expect a summed output. Delays aren't a problem either because we are assuming a time-invariant system.

Despite the intuitive appeal of this approach, there are more powerful ways of representing a signal—representing a complex signal as a sum of different frequency, amplitude and phase sinusoids. In fact, we will find that *any* input signal can be represented as the sum of sinusoids. Assuming the system is linear and time-invariant, we can use the Bode plot to find how each input sinewave will be transformed by having its magnitude and phase altered.

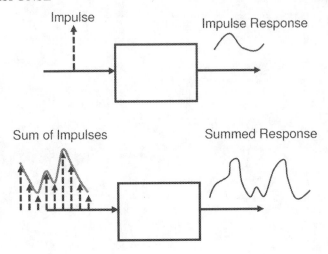

Figure 8.15: Using the impulse response to predict the response to a sum of impulses.

8.6.1   CONVOLUTION

Consider what the above means by thinking about a doctor performing a knee jerk reflex test. Although it is not a perfect mathematical impulse, the mechanical "smack" of the hammer is the input and the output is the angle of the knee over time. Since we know that the response in the s-domain is $H(s)$, the response we observe in the Time Domain must be $h(t)$. If we have an analytical function for $h(t)$, finding $H(s)$ is a "simple" matter of taking the inverse Laplace Transform. Once we have that, we could figure out any combination of impulses to the knee.

But in a real situation we will have some messy time course for $h(t)$. And here we confront an issue—in a real experiment our input, $\delta(t)$, and output, $h(t)$, will be in the Time Domain. In theory we could just take the inverse Laplace Transform, but that would get very ugly for a time course from a real experiment.

So we might then want to try to do what we did for the simple systems above—find the poles and approximate our system. Again that might work. But biological systems (and devices that interface with biological systems) can be very complex and we may not want to find an exact set of differential equations for our system. Our fundamental problem is that it is easy to experimentally get $h(t)$, but it is $H(s)$ that we need to start predicting input-output pairs. This all falls out from the following relationships

$$H(s) = \frac{o(s)}{i(s)} \tag{8.55}$$

$$h(t) \neq \frac{y(t)}{x(t)} \tag{8.56}$$

Luckily there is a relationship between $h(t)$, $x(t)$ and $y(t)$ called the *convolution*. You will be spared a formal derivation of the convolution, but below is the relationship.

$$y(t) = \int_{-\infty}^{\infty} x(\tau)h(t - \tau)d\tau \qquad (8.57)$$

From left to right, the output, $y(t)$, is equal to the sum of the input, $x(\tau)$, and infinitely many *shifted* $(t - \tau)$ versions of the impulse response, $h(t)$. Note here that you can think of the input $x(\tau)$ as scaling a bunch of impulse responses. Also important is that the integral is *not* over time, but the time shift. So we are in fact doing what was shown in Figure 8.15. Convolution is a way to get $y(t)$ from $x(t)$ and $h(t)$ without needing to convert everything into the s-domain.

Notice that in Equation 8.57 we would need some analytical function for both $x(t)$ and $h(t)$ to come up with an analytical value of $y(t)$. Although this might make for a nice home-work problem, the continuous version is not useful. Instead, we can think of a discrete version of convolution as

$$y[n] = \sum_{k=-\infty}^{\infty} x[k]h[n - k] \qquad (8.58)$$

The primary reason to think about convolution discretely is because it is how a computer would implement convolution. Another reason to consider the discrete version is that in some ways it is easier to visually understand what is happening.

We can think about how we might implement Equation 8.58. First, we flip $h[n]$ about the y-axis. Second, shift this version of h to the *right* by n. Third, multiply this flipped and shifted version of $h[n]$ with $x[n]$ at each value and sum the result. This will give you a single number for $y[n]$ at n (remember this is the amount you shifted $h[n]$ to the right). This only gives you the value of y at n (so $y[n]$ for that value of n). Lastly, you need to repeat these steps for each value of n. How do you know when to stop? In reality, your known signals ($x[n]$ and $h[n]$) will be of a finite length. If you keep shifting and multiplying, eventually you will slide the signals right past each other, after which all the summations will be zero. To gain some insight, the idea is shown graphically in Figures 8.16 and 8.17.

And then we would follow the steps below

- $y[n]$ for $n < 2$ = no overlap = 0

- $y[2]$ = shift right by 2 = (3)(1) = 3

- $y[3] = (3)(2) + (2)(1) = 8$

- $y[4] = (3)(-1) + (2)(2) + (1)(1) = 2$

- $y[5] = (2)(-1) + (1)(2) = 0$

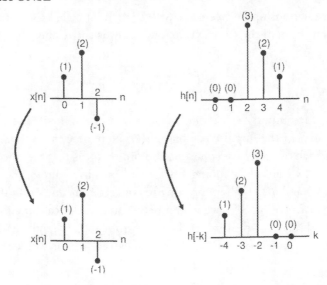

Figure 8.16: A graphical representation of the first step in the convolution of $x[n]$ and $h[n]$.

- $y[6] = (1)(-1) = -1$
- $y[n] =$ for $n > 6 = 0$

And so our final result is shown in Figure 8.17.

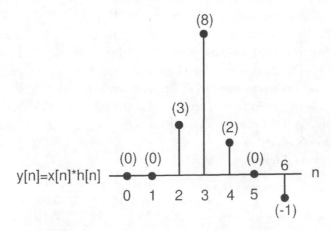

Figure 8.17: A final graphical representation of the convolution of $x[n]$ and $h[n]$ in Figure 8.16.

Rather than write out the entire integral (or summation) the operation of convolution is often written using the $*$ notation.

$$y(t) = x(t) * h(t) \tag{8.59}$$

So, in summary

$$y(t) = x(t) * h(t) \tag{8.60}$$
$$H(s) \times X(s) = Y(s) \tag{8.61}$$

or convolution in the Time Domain is the same as multiplication in the s-domain.

8.6.2   DECONVOLUTION

But what do we do if we know $h(t)$ and $y(t)$ but not $x(t)$? This may seem strange; don't we always know the input $x(t)$? Actually we don't always know it because remember that $x(t)$ might be an internal signal (i.e., it might arise from some other subsystem inside of a larger system). This problem comes up a lot in a number of medical situations and we will discuss one in detail below.

To find $x(t)$ we essentially need to "undo" convolution. This is achieved using *deconvolution*

$$Y(s) = H(s) \times X(s) \tag{8.62}$$
$$y(t) = x(t) * h(t) \tag{8.63}$$

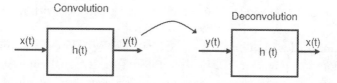

Figure 8.18: A graphical representation of deconvolution.

Now consider the entire system, where we input $x(t)$, convolve it with $h(t)$, but then deconvolve with $h'(t)$, to get back to $x(t)$, as in Figure 8.18. If you think about it, the input and output are $x(t)$ and only one transfer function will do that, $h(t) * h'(t) = \delta(t)$ or in the s-domain $H(s)H'(s) = 1$. So as is in the true definition of an inverse, $h'(t)$ must undo $h(t)$.

The meaning of an inverse is much easier to think about in the s-domain because it is simply $H'(s) = 1/H(s)$. What this means is that zeros become poles, and poles become zeros. This poses an interesting problem. What if $H(s)$ is stable (poles all in the left half plane), but it has zeros in the right half plane? When we form $H'(s)$ it is going to have its poles in the right half plane and therefore be unstable. There are ways to deal with this, but they are beyond the scope of this text.

8.7   OTHER APPLICATIONS

There are some more advanced applications of the concept of an impulse function that are important in biomedical engineering. We have only considered if the impulse function is an impulse in time. However some systems vary in space and time. Here we must define the impulse as

$$\delta(t, x) = \begin{cases} 1 & \text{if } |t| \leq \frac{t}{2} \text{ and } |x| \leq \frac{x}{2} \\ 0 & \text{Otherwise} \end{cases}$$

A physical interpretation of this would be in the pluck of a string. The impulse occurs at a very specific place on the string and at a very specific time. Once the pluck is over, the string will vibrate. In fact, it is the vibration, and decay to steady state, that would be the impulse response in this case. This idea is useful in thinking about the electrical stimulation of a nerve axon, the targeting of radiation to a tumor and ultrasonic treatment of injured joints.

Another application is in imaging of biological tissue. A basic block diagram of a medical imaging technology is shown in the top of Figure 8.19. Here some energy is generated (e.g., photons, X-rays, sound) by an emitter and then passed though the body. The body will change these energetic particles (or waves) in some way. Then there is a detecting device that must record how they have changed. The key is that we are after something about the human body as a system, $H_b(x, y)$. The problem is that we know the input from the emitter $c(x, y)$ and the output of the detector, $e(x, y)$, but we don't know the signals or systems in between. So the real problem is that we need to somehow first backtrack to find the signal $d(x, y)$ so we can get to the body system, $H_b(x, y)$. The reason we keep using (x, y) is because all of this is assumed to be creating some sort of an image. If it was an image that changes over time (as is often the case), we would need to add (x, y, t).

The key is to first find the transfer function of the detector with no body present. In the bottom part of Figure 8.19, we have made the signal $d(x, y)$ an impulse. But because we are talking about an image, this impulse is a dot (a spatial impulse). We can then find how the detector transforms that dot into an image, $e(x, y)$. This output from a spatial impulse, known as the *Point Spread Function*, is used to derive a transfer function, $h_d(x, y)$, for a medical detector. Now that we know the device impulse response, $h_d(x, y)$, we can record an image from a real human, $e(x, y)$, and then find $d(x, y)$. You might see that this is just a deconvolution (we know the output and system and want to find the input). We will not go through the details, but we now have $c(x, y)$ (input to the human body) and $d(x, y)$ (output from the human body) and can begin to learn about $H_b(x, y)$. Again, the goal here is not to go through all of the details of medical imaging. But, in a nutshell, interpreting the pictures that come out of a medical imaging device requires a series of transforms.

8.7.1   OTHER USEFUL TEST SIGNALS

There are a number of other useful test signals that have applications in the biomedical field. A *ramp* function is one that slowly grows from 0 to some value over a long period of time. You can

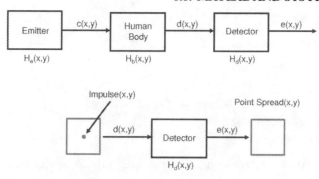

Figure 8.19: A block diagram of how most medical imaging devices work.

think of it as the sum of many increasing amplitude impulses. Another function with a similar idea is the *chirp*, so called because when played on speakers it sounds a little like the chirp of a bird. It is a signal that starts out with a low frequency sinewave but over time the frequency gets higher and higher. We will use the chirp signal in Chapter 12 to test the filters we build. The last signal is *noise*! It may seem strange to send noise into a system as a test signal, but it can be useful for a number of reasons. For example, we might want to know if the system can respond in some recognizable way to an impulse when it is embedded in noise. We also might want to know how a subsystem will handle background noise from other processes once we integrate it into a larger system. Lastly, there are times when we want to test out an algorithm for finding features in a signal that contains noise. We will learn more about noise in later chapters, but for now it is worth knowing that engineers talk about many different types of noise—*white noise*, *pink noise*, *colored noise*, *Gaussian noise*, *random noise*. In reality these overlap with one another and there is an art to figuring out which type of noise is best to introduce.

8.8   MATLAB AND SYSTEM RESPONSES

Matlab has a number of commands that are very helpful in creating the various responses discussed in this chapter. For example, let's consider the Low Pass RC circuit from Figure 8.1. We can first create the transfer function

```
>> R = 10000; %in Ohms
>> C = 2e-7; %in Farads
>> Num = [1];
>> Den = [R*C 1];
>> GL = tf(Num,Den);
>> impulse(GL);
```

We can then use the command *impulse* to create the impulse response for this circuit.

```
>> impulse(GL);
```

To make it look a bit nicer we can get the actual signal

```
>> [y,t] = impulse(GL);
>> plot(t,y);
>> title('Impulse Response for Low Pass RC');
>> xlabel('Time (sec)');
>> ylabel('Signal');
```

You might notice that the *impulse* command automatically computes how much time should be plotted. You can change this default by specifying a *timevector* and then passing it into the *impulse* function as the second argument.

```
>> dt = 0.01;
>> timevector = 0:dt:10;
>> impulse(GL,timevector);
```

To get the step response, Matlab has the *step* command.

```
>> [y,t] = step(GL);
```

And as you may have suspected, you can create a Bode plot with the command *bode*

```
>> bode(GL);
```

The last command, *lsim*, is to find the response of any arbitrary input. For the sake of interpretation, we will assume that our arbitrary signal is a sum of sinusoids.

```
>> t = 0:0.0001:1;
>> f = [100 204 306];
>> V = 1.5*\cos(2*pi*f(1)*t) + 1.75*sin(2*pi*f(2)*t) + 2.0*sin(2*pi*f(3)*t);
```

Here we have created a signal, V, that is composed of three sinewaves with frequencies 100, 204 and 306 Hz. To find out the response of our RC circuit to this signal

```
>> [Out,t]=lsim(GL,V,t);
>> plot(t,V,'r--');
>> hold on
>> plot(t,Out);
```

The *lsim* command takes in the transfer function, GL, as well as the time vector and input signal, V. It then sends out the response, Out, along with the time vector. The last three lines are simply plotting the input (red dashed line) and output. Note that to get the responses to the highpass RC, and our example second-order system discussed in the chapter, all you would need to do is make a new transfer function and then issue the same *impulse*, *step* and *bode* commands to generate the figures in this chapter. It is also worth noting that there are a number of other commands that are useful in obtaining systems responses, for example *conv* and *deconv*, that won't be covered here.

8.9 EXERCISES

1. To characterize the overall performance of the auditory system, clinicians present pure sinusoidal tones (usually to one ear at a time) to determine the signal amplitude at which the sound is just audible to the subject. To a first approximation, the human auditory system can be modeled as a first-order RC circuit. The Bode plot is shown in Figure 8.20.

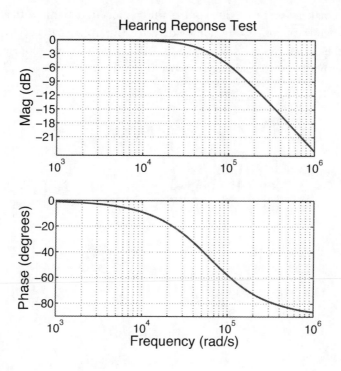

Figure 8.20: A Bode plot for a hearing test.

    a. Does the auditory system simulate a lowpass or highpass circuit?

    b. Find the cut-off frequency in Hz. Explain your method.

    c. Sketch a 1 Volt peak-to-peak sinewave with a frequency at the cut-off frequency. On the same plot, sketch the auditory system response. Explain your answer.

2. The Schmid-Schoebein and Fung model for respiration can be modeled as

$$\frac{\epsilon(s)}{p(s)} = \frac{0.04s^3 + 0.1}{-0.0035s^3 + 2.4s^2 + 32.5s}$$

where $p(t)$ is an applied pressure and $\epsilon(t)$ is a fractional change in lung wall motion.

a. Derive an analytical expression for the impulse response in the s-domain.

b. Use Matlab to show the impulse response in the Time Domain. Show your commands.

c. Derive an analytical expression for the step response in the s-domain.

d. Use Matlab to show the step response in the Time Domain.

3. Neural pacemakers can be used to control the tremors in patients with advanced Parkinson's disease. When the signs of a tremor are detected (input) a small current waveform (output) is sent to the neural tissue. It is known that this device can be modeled as a second-order system and that the step response is shown in the Figure 8.21.

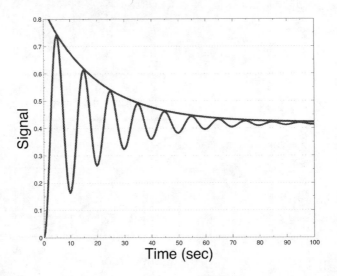

Figure 8.21: A step response for a neural pacemaker.

a. Estimate the steady-state.

b. Estimate the overshoot.

c. Estimate the undershoot.

d. Estimate the natural frequency in rad/sec. Explain your method.

e. Estimate the decay of the oscillations.

f. Estimate the pole locations.

4. A major limitation of pH electrodes is their slow response time. A typical "fast" commercially available pH electrode has a time constant of two seconds and can be modeled as a first-order lowpass RC circuit.

a. Derive an analytic expression for the transfer function in the s-domain for this electrode. Note that this transfer function is defined as the measured pH (output) divided by the actual pH (input).

b. Create a Bode plot for the transfer function.

c. In arterial blood, the primary oscillatory component of pH is due to respiration. A rough estimation of a respiration rate is 0.1Hz. Plot how the pH transfer function will respond to respiration and comment on how important respiration may be to pH measurement.

d. Proper functioning of neurons in the brain depended upon pH. Brain tissue pH has several oscillatory components. Assume that the actual pH is defined by

$$0.01 \sin(2\pi 0.225t) + 0.002 \sin(2\pi 4t) + 0.005 \sin(2\pi 0.1t) \tag{8.64}$$

Plot the actual recording from the pH sensor over a time course which shows representative behavior. Which component of the signal do you expect to have significant errors? Justify your answer.

5. Iron is an important element that is necessary for the production of hemoglobin, myoglobin, and cytochromes. The pathways by which dietary iron is distributed in the blood plasma is complicated. For this reason, the response of the body to an iron input was mapped out in mice using bolus injections of an iron tracer (56FE). This mapping yielded the following information:

- There are system zeros at -3 and $3 \pm 2j$
- There are system poles at $\pm 2j$ and a double pole at -2

a. Derive the transfer function for this system.

b. Derive an analytic expression for the step response. Note that you do not need to simplify your expression.

c. Find the step response for this system using Matlab. *Note:* It may be helpful to use the series command to simplify the expression in part b.

6. A suspended muscle is often modeled as a mass, spring and dashpot as in the Figure 8.22. We can sum forces to arrive at

$$F(t) = M\frac{d^2x}{dt^2} + R\frac{dx}{dt} + Kx$$

or in the Laplace Domain

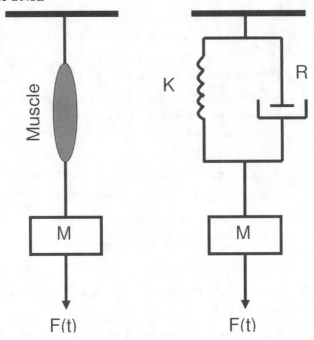

Figure 8.22: Model of a suspended muscle.

$$F(s) = Mx(s)s^2 + Rx(s)s + Kx(s)$$
$$F(s) = x(s)\left[Ms^2 + Rs + K\right]$$

and so the transfer function is

$$\frac{x(s)}{F(s)} = \frac{1}{Ms^2 + Rs + K}$$

assuming that all initial conditions are zero. We can then transform our transfer function into the standard form for a second-order transfer function

$$\frac{x(s)}{F(s)} = \frac{\frac{1}{M}}{s^2 + \frac{R}{M}s + \frac{K}{M}}$$

$$\frac{x(s)}{F(s)} = \frac{\frac{1}{M}}{s^2 + 2\delta\omega_n s + \omega_n^2}$$

where

$$\omega_n = \sqrt{\frac{K}{M}}$$

$$\delta = \frac{R}{2\sqrt{M}\sqrt{K}}$$

a. If the mass is assumed to be 1 gram and the natural frequency is 1 Hz, find the constant for the spring. Be sure to report your units.

b. Explain how the dashpot (R) of the muscle may lead to undamped, underdamped and overdamped systems. Give a criteria for the value of R that would transition between these different types of behavior.

7. Given the following poles from a previous homework problem

   a. $-3, -5$

   b. $-1, -1 \pm -2j$

   c. $\pm 3j$

   d. $-4, 2, \pm 3j$

show the impulse response and discuss the difference between these systems and a base system that has poles as $-1, -1 \pm j$. You should compare any oscillations, decay or growth.

8. A simple model of postural control is a mass on an inverted pendulum. The model assumes the entire body mass, M, is placed at the center of gravity and balanced atop a straight "stick" leg, a distance, L, from the floor. T is the effective torque resulting when the person is off balance (as occurs during walking). There are three torque components: 1) due to circular acceleration of the body mass, 2) due to rotational resistance, R at the ankle joint, and 3) due to the component of gravitational forces perpendicular to the leg. If we assume that for small angles $\sin(\theta) = \theta$, then:

$$T(t) = ML^2 \frac{d^2\theta}{dt^2} + R\frac{d\theta}{dt} + MgL\theta(t)$$

a. Find the transform $\frac{\theta(s)}{T(s)}$. You may assume all initial conditions are zero.

b. Find the locations of the poles as a function of R, M, L and g.

c. If there is no rotational resistance, draw the pole locations in the s-domain and explain the Torque response in time.

d. In a more realistic case $R > 0$. Based upon pole locations in the s-domain, discuss the two possible responses of this system.

CHAPTER 9

Control

Most biological systems and devices have some sort of feedback to either stabilize the system or to correct for errors, or both. In Chapter 7 we saw how placing a very simple feedback element (with some gain K) could generate a large variety of system responses (from stable to unstable to oscillatory).

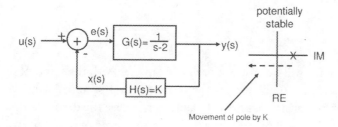

Figure 9.1: An inherently unstable system made stable by feedback, from Chapter 7.

You might imagine that your body uses this type of feedback to help regulate temperature (stable), fight off a nasty infection (unstable) or generate your daily cycle of blood glucose levels (oscillation). But how does your body *know* what to make K? In fact, your body and many other systems have a smart way to figure out how to control a system. Of course your body, and other complex systems, have some very intricate and robust methods of control. In this chapter we introduce the idea of system control. Understanding the more complex system control is beyond the range of this text.

9.1   THE GENERIC CONTROL MODEL

There are many ways that a system can be controlled using feedback and some additional sub-systems, however, the generic model is shown in Figure 9.2.

First you should notice that Figure 9.1 and Figure 9.2 are not the same and that the controller, $G_c(s)$, is not part of the feedback loop where we might expect it to be. Below we will examine each sub-system and signal of Figure 9.2.

$G_p(s)$ is the *plant* system that is going to be controlled. More specifically, there is some variable of the system (usually the output, c) which we want to be at some desired value. Unlike many of our previous systems, the input, r, is not some arbitrary value. It is the *desired* value we wish the output, c, to be and is usually called the *setpoint*. Most often, the engineer has control

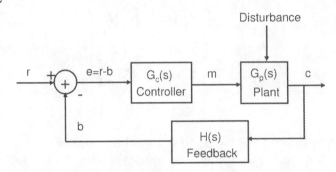

Figure 9.2: A generic controller showing the Plant system to be controlled, the Feedback system and the Control system.

over what the setpoint will be. You are probably familiar with the setpoint on your thermostat—it is what you wish the temperature of the room to be. The term "plant" comes from chemical engineering where the plant is a physical plant. In this case the desired output would be some set mix of chemicals that have a property that can be reproduced over and over again. It is also worth nothing that there is a "disturbance" signal coming into the plant. This is any signal from the outside that is not under the control of the engineer. We will see some examples of disturbance signals later in the chapter.

$H(s)$ is a *feedback transfer function*. The job of the feedback sub-system is often to simply measure the actual output, c. In this way, the actual output can be compared to the desired output, r. But in making the measurement we might need to change the actual output in some way to a signal b. You can think of $H(s)$ as a conversion. Often this is because the output of our system is being measured by a real device (e.g., transducer). As such, the measured signal, c, may need some filtering, amplification, calibration, or unit conversion, after which we have the signal b. It is $H(s)$ which will perform all of these functions.

The comparison of the measured output and desired output is performed by a junction that subtracts b from r. This generates an *error* signal, e. Sometimes, e is called the *actuating signal*.

$G_c(s)$ is the controller and may take many forms. The basic idea is that the controller will use the error signal, e, to generate a "nudge" (i.e., a small correction signal). This nudge, m, to the plant system is in effect a second input (in addition to the disturbance), that will send the actual output closer to the desired output. Therefore, the goal of the controller is to make a nudge to the system so that $e = 0$.

In most systems, the plant is already determined and the feedback system is in place simply to generate an error signal. So all of the design of a controller is contained within $G_c(s)$.

If that was all there was to the picture, we might imagine designing a very particular controller $G_c(s)$ for a particular $G_p(s)$. It might take some tuning of $G_c(s)$, but eventually we might imagine getting it right and then never needing to change the controller again. This would be very similar to the idea of tuning K in Figure 9.1.

Nearly all of the systems that we engineers must deal with, however, are *open* systems. They interact in some way with their environment, taking inputs from the environment and sending outputs to the environment. This idea is represented by the *disturbance* signal.

The fact that outside disturbances exist in fact force the idea of an adaptive controller. My grandmother was an expert breadmaker. But she had no recipe. In fact, she couldn't read. She only understood the basic ideas of counting and didn't have any measuring cups other than her hands. But her bread was always wonderful. When my mother asked for the recipe, my grandmother laughed and said that she couldn't do it. So my mother simply watched and measured everything that my grandmother put into the bread, writing down all of the steps. She then went back home and tried to make the bread. It was okay, but not the same. When she asked why it didn't come out the same, my grandmother laughed again and said that she didn't follow a recipe because the recipe wasn't what was important, it was the feel and consistency of the bread. She was feeling if it had the right mix of ingredients, and if it didn't she would add what was missing until it was right. She then went on to explain that making bread in America was a little different than making bread in Italy because the flour was different. And that it was different at different times during the year because of the humidity. And that it also depended on how big a batch she was making. There were so many variables that they could not all be accounted for, so instead she simply measured the one variable (consistency) that mattered.

The example of making bread may not seem like engineering, but making concrete is not all that different—it requires the right consistency. And that consistency depends on a number of environmental factors. And it is very important to measure the right variable. In another example, brewing beer, the pH of the wort is important in having healthy yeast and flavor extraction. In a big brewery, this is carefully monitored. If the pH is off, it will be adjusted. The same is done for color, and flavor compounds, all to get the beer to be the same in every batch, regardless of disturbances like time of year, variation in hop or barley harvest or water quality.

9.2 EVALUATING A CONTROLLED RESPONSE

Before we jump into types of controllers, we will discuss how to evaluate the time and frequency domain response of a controller. One very important point is that the control system, $G_c(s)$, is always evaluated as it functions within the entire controlled system (all of Figure 9.2), not on its own. In other words, we are not interested in evaluating $\frac{m(s)}{e(s)}$, but $\frac{c(s)}{r(s)}$.

9.2.1 TIME DOMAIN EVALUATION

The Time Domain properties of a controller are evaluated using a step response to the entire closed loop control system. Making r(t) equal to a step function is the same as saying that we are going to suddenly change the setpoint. The result is that the controller will attempt to respond and make $c(t) = r(t)$ again. We will measure how well the controller does by how well the output approximates a step function.

Since we are using a step function, the Time Domain evaluation is the same as presented in Figure 9.3. So we can define steady-state, overshoot/undershoot, rise time and settling time. The additional, and very important, measurement is the *Steady-State Error* which is defined as the percentage difference between the steady-state (actual output) and setpoint (desired output).

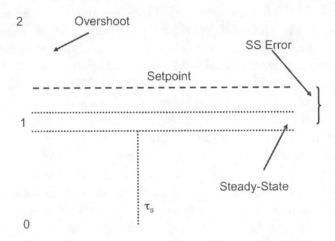

Figure 9.3: Quantify the time-domain step response of a generic controller.

9.2.2 FREQUENCY DOMAIN EVALUATION

A side effect of many controllers is that they act as filters in the frequency domain. The concept of the frequency domain was first encountered in a discussion on Bode plots. We will use the top half of the Bode plot to characterize which frequencies a controller will naturally amplify and which frequencies will naturally be attenuated. The bottom half of the Bode plot will tell us something about the delays (phase shifts) that may be introduced.

9.3 ON-OFF CONTROLLERS

Consider what happens to your system if the output is right at the setpoint. The error signal will be zero, which should mean that the controller does nothing (i.e., no nudge). Now consider what happens if the error is positive. Here we would want to send a small positive m signal into the plant. Likewise, if the error is negative we would want to send a small negative m signal into the plant. Mathematically,

$$m(t) = \begin{cases} m_{max} & \text{if } e > 0 \\ 0 & \text{if } e = 0 \\ m_{min} & \text{if } e < 0 \end{cases}$$

or graphically

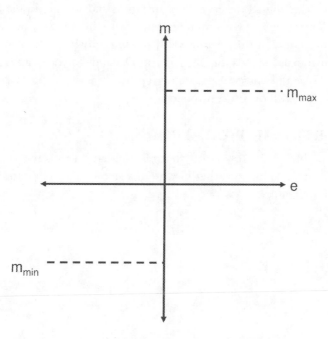

Figure 9.4: The response, m, of an on-off controller to an error signal, e.

Figure 9.4 requires a bit of explanation. It is a plot of the output of the controller, m, given some error input, e. These types of plots of e versus m can be helpful in thinking about different types of controllers.

The advantage of the on-off controller is that we only need to chose m_{min} and m_{max}. The only other complication is that in some systems, a positive error will require a negative m signal, and the opposite for a negative error. This is easily fixed by switching m_{min} and m_{max}.

You may notice that the on-off controller is not really on or off. In fact, it is usually sending some m signal all the time, either m_{min} or m_{max}. You can think of this as a temperature-controlled room where m_{min} is when the air conditioning is turned on all the way. Likewise, m_{max} is when the heat is turned on all the way. A bit of thought will show that if both the air conditioner and heater are in use and both have the same setpoint, the system will oscillate around the setpoint. First the heater will come on and warm up the room above the setpoint. Then the air conditioning will turn on and cool the room below the setpoint. This cycle will continue because an on-off controller will overreact to even small errors. For this reason most on-off controllers set either m_{min} or m_{max}

to zero (consider what this does to Figure 9.4). The result for the temperature-controlled room is that we will use either the heating controller or cooling controller, but not both.

9.4   PID CONTROLLERS

On-off controllers are good in a variety of circumstances and are the simplest to understand. But there are some major problems if we want to reach our setpoint, reach it quickly and do so without overshoot and undershoot. In this section we will build up what is known as a *PID* controller. P stands for Proportional, I stands for Integral and D stands for Derivative. These are by far the most popular controllers in engineering, taking up nearly 90% of the market. We will start by fixing a major problem with on-off controllers.

9.4.1   PROPORTIONAL (P) CONTROL

The fundamental problem with the on-off controller is that it reacts the same way to both small and large errors. To fix this, imagine that instead of a step function in Figure 9.4, we use a more linear function, as in Figure 9.5.

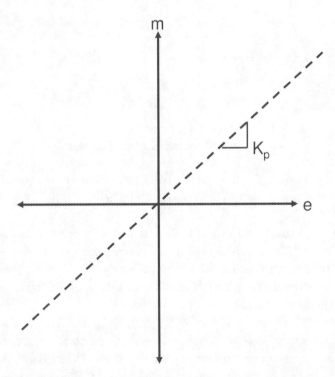

Figure 9.5: The response, m, of a proportional controller to an error signal, e. The slope is given as the value K_p.

The idea here is that if the error, e, is large, the nudge, m, should also be large. If the error is small, the nudge should be small. Again, if $e = 0$ then $m = 0$. In other words, m should be *proportional* to e or:

$$m(t) = K_p e(t) \tag{9.1}$$
$$m(s) = K_p e(s) \tag{9.2}$$

Given this relationship the transfer function for the controller will be:

$$G_c(s) = \frac{m(s)}{e(s)} = K_p \tag{9.3}$$

One common modification to the proportional controller is to use the formulation in Equation 9.3 for errors that are within a certain range. However, outside that range we clip the $m(t)$ to be either m_{min} or m_{max}. Figure 9.6 shows an example of a linear region (slope controlled by K_p) with $m(t)$ clipped for large errors.

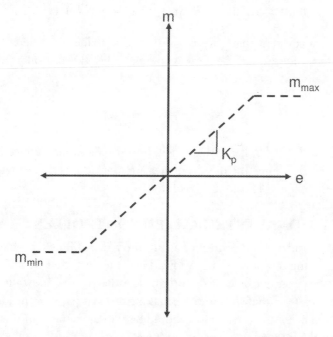

Figure 9.6: A proportional controller where the response has been clipped.

9.4.2 PROPORTIONAL DERIVATIVE (PD) CONTROLLER

Consider what happens if the system is far away from the setpoint and is being controlled by a proportional controller. If the proportional constant K_p is small (i.e., gentle slope), then the error will be reduced very slowly. Alternatively, if K_p is large then the error may drastically overshoot and undershoot the setpoint.

To help both cases it would be useful to know how quickly the system is responding to the m signals. In other words, we want the controller to make a nudge and then record how much the error changed. To detect this change we will need to give one nudge and then check to see how much difference it made. The next nudge will then be altered appropriately. What we need is some way to evaluate how much of a difference was made in the error. A bit of thought will reveal this change is the slope of $e(t)$ or $\frac{de}{dt}$. The idea is that knowing the derivative of e(t) will help *anticipate* where the error will be in the future (if the same nudge was given again) and a correction term can be added to the proportional controller. This correction term is simply the *derivative* of $e(t)$

$$m(t) = K_p e(t) + K_d \frac{de}{dt} \tag{9.4}$$

$$m(s) = K_p e(s) + K_d s e(s) = (K_p + K_d s) e(s) \tag{9.5}$$

where K_d is simply a constant that governs how large to make the differential correction term. You should note that using Equation 9.5, we can define the PD controller transfer function as:

$$G_c = \frac{m(s)}{e(s)} = K_p + K_d s \tag{9.6}$$

The advantage of the PD controller is that it will improve damping, reduce the maximum overshoot and reduce the rise time. The disadvantages are that little is done about steady-state error and the transfer function acts like a highpass filter.

9.4.3 PROPORTIONAL INTEGRAL (PI) CONTROLLER

If we back up and consider the P controller again, there can still be a problem where the actual output never reaches the setpoint. What we need is some term that will take into account how long the error has been on one side of the origin. In other words, if $e(t)$ is always negative we should add a correction term that will continue making $m(t)$ larger. And the longer $e(t)$ is negative (or positive) the larger this correction term should be. What we need in this case is the integral of $e(t)$. The reason the integral works is because we are simply adding up a history of what the error has been in the past. This integral is like building up momentum and is sometimes called *Wind-Up*. It helps to keep track of past errors.

Mathematically a PI controller has the form

$$m(t) = K_p e(t) + K_i \int e(t)dt \tag{9.7}$$

$$m(s) = K_p e(s) + \frac{K_i}{s} e(s) = [K_p + \frac{K_i}{s}]e(s) \tag{9.8}$$

Where K_i is simply a constant that governs how large to make the integral correction term. The transfer function for the PI controller is therefore

$$G_c = \frac{m(s)}{e(s)} = K_p + \frac{K_i}{s} \tag{9.9}$$

In general, the advantage of the PI controller is that it will improve steady-state error and dampen the maximum overshoot. However, the PI controller acts as a lowpass filter and may decrease rise time.

9.4.4 PROPORTIONAL INTEGRAL DERIVATIVE (PID) CONTROLLER

As there are advantages of adding both derivative and integral correction terms, we can add both to the proportional controller to arrive at the PID controller.

$$m(t) = K_p e(t) + K_d \frac{de}{dt} + K_i \int e(t)dt \tag{9.10}$$

$$m(s) = K_p E(s) + K_d s e(s) + \frac{K_i}{s} e(s) = [K_p + K_d s + \frac{K_i}{s}]e(s) \tag{9.11}$$

$$m(s) = [K_d s^2 + K_p s + K_i]e(s) \tag{9.12}$$

And so the transfer function for a PID controller is

$$G_c(s) = [K_d s^2 + K_p s + K_i] \tag{9.13}$$

Here we can think of these terms as reacting to future (K_d), present (K_p) and past (K_i) errors.

9.4.5 CHOOSING CONSTANTS

By picking the constants appropriately, we get the best of the P, PI and PD controllers. But this is not always a simple matter. In fact, finding good values of K_p, K_i and K_d is called *tuning* and is typically an iterative process:

1. Obtain the open-loop response and determine what needs to be improved.

2. Add P control to improve rise time.

3. Add D control to improve over and undershoot.

4. Add I control to improve steady-state.

5. Adjust K_p, K_i and K_d until desired response is found.

Two additional warnings are in order. First, for many controllers there is no need to use all three terms. It is best to keep your controller as simple as possible while still obtaining the desired response. So keeping a controller either P, or PI or PD will make tuning much easier. Second, the proportional, integral and derivative terms are somewhat dependent upon one another. So in adjusting one parameter you may inadvertently make an undesirable change.

Although there is no sure-fire algorithm for tuning the P, I and D constants some have been developed that follow a version of the iterative process above.

9.4.6   ALTERNATIVE FORMULATION

Sometimes a PID controller is formulated a bit differently by defining:

$$T_i = \frac{K_p}{K_i} \tag{9.14}$$

$$T_d = \frac{K_d}{K_p} \tag{9.15}$$

Where T_i is called the *Integral Rate Constant* and T_d is the *Derivative Rate Constant*. Given these definitions:

$$m(s) = K_p[1 + \frac{1}{T_i s} + T_d s]e(s) \tag{9.16}$$

$$m(t) = K_p[e(t) + \frac{1}{T_i} \int e(t)dt + T_d \frac{de(t)}{dt}] \tag{9.17}$$

The advantage of this formulation is that T_i is a measure of how far the I correction term will look into the past. Likewise, the T_d term is a measure of how far the D correction term will project into the future. They are not absolute values (they don't have units of time), but relative, and can help in tuning values.

9.5   EXAMPLE OF A PID CONTROLLED SYSTEM

Because we cannot evaluate a controller on its own, we will consider controlling the second-order plant system

$$G_{open}(s) = \frac{1}{s^2 + 10s + 20} \tag{9.18}$$

It is always good practice to first evaluate the open-loop response. If a step function is used as the input we are basically changing the setpoint from 0 to 1. So the ideal response of our system would be for the plant output to jump from 0 to 1. The step response for the system is shown in Figure 9.18.

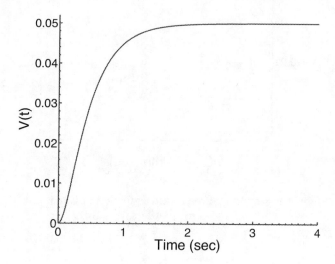

Figure 9.7: Open loop step response of the transfer function in Equation 9.18.

The steady-state value is 0.05 so the steady-state error is 95%! As we are way off, the next step is to determine the closed-loop behavior by setting $G_c(s) = 1$ and $H(s) = 1$.

$$G_{closed}(s) = \frac{G_p(s)}{1 + H(s)G_p(s)} \tag{9.19}$$

$$G_{closed}(s) = \frac{\frac{1}{s^2+10s+20}}{1 + \frac{1}{s^2+10s+20}} \tag{9.20}$$

$$G_{closed}(s) = \frac{1}{s^2 + 10s + 21} \tag{9.21}$$

The response in Figure 9.8 does not change significantly. Note, however, that the closed loop response for some systems will drastically change the output. It is always good to check.

Next we can include a controller. Note that $G_c(s)$ and $G_p(s)$ are in series, so:

$$G_{control}(s) = \frac{G_p(s)G_c(s)}{1 + H(s)G_p(s)G_c(s)} \tag{9.22}$$

If we assume that the control is the full PID implementation, then:

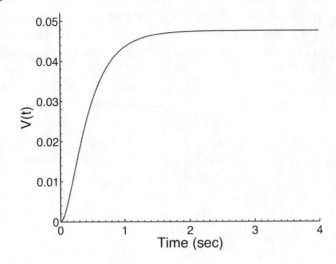

Figure 9.8: Closed loop step response of the transfer function in Equation 9.21.

$$G_{control}(s) = \frac{G_p(s)[K_p + \frac{K_i}{s} + K_d s]}{1 + H(s)G_p(s)[K_p + \frac{K_i}{s} + K_d s]} \tag{9.23}$$

By setting the appropriate constants to zero, we can now use Equation 9.23 to examine the effect of a P, PD, PI and PID controller. For example, consider a P controlled system by setting $K_i = 0$ and $K_d = 0$ and $K_p = 300$:

$$G_{control}(s) = \frac{G_p(s)K_p}{1 + H(s)G_p(s)K_p} \tag{9.24}$$

$$G_{control}(s) = \frac{K_p}{s^2 + 10s + (20 + K_p)} \tag{9.25}$$

$$G_{control}(s) = \frac{300}{s^2 + 10s + (20 + 320)} \tag{9.26}$$

Notice that we have improved the steady-state error (although it is not exactly at the desired value of one) but now there are oscillations (undershoot and overshoot) as we reach the steady-state.

To reduce these oscillations we can consider a PD controller ($K_i = 0$ and $K_d = 10$ and $K_p = 300$). Again using Equation 9.23

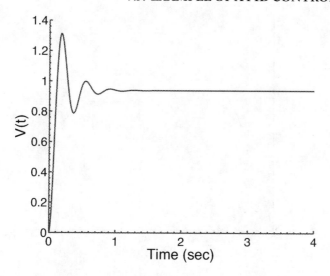

Figure 9.9: A step response of the transfer function in Equation 9.23 with proportional control added.

$$G_{control}(s) = \frac{G_p(s)[K_p + K_d s]}{1 + H(s)G_p(s)[K_p + K_d s]} \tag{9.27}$$

$$G_{control}(s) = \frac{K_d s + K_p}{s^2 + (10 + K_d)s + (20 + K_p)} \tag{9.28}$$

$$G_{control}(s) = \frac{10s + 300}{s^2 + 20s + (20 + 320)} \tag{9.29}$$

Notice in Figure 9.10 that compared to the P controller (also note that the time axis has been changed) the overshoot and settling time have been reduced (i.e., less violent oscillations), however, little has been done to improve the rise time or steady-state error.

Next we can consider a PI controller ($K_i = 70$ and $K_d = 0$ and $K_p = 30$) with the resulting response shown in Figure 9.11. Using Equation 9.23

$$G_{control}(s) = \frac{G_p(s)[K_p + \frac{K_i}{s}]}{1 + H(s)G_p(s)[K_p + \frac{K_i}{s}]} \tag{9.30}$$

$$G_{control}(s) = \frac{K_p s + K_i}{s^3 + 10s^2 + (20 + K_p)s + K_i} \tag{9.31}$$

$$G_{control}(s) = \frac{30s + 70}{s^3 + 10s^2 + 50s + 70} \tag{9.32}$$

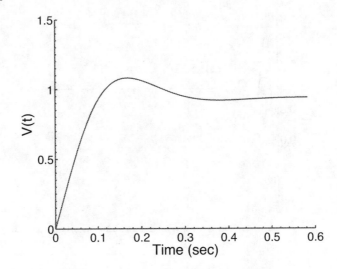

Figure 9.10: A step response of the transfer function in Equation 9.23 with proportional-derivative control added.

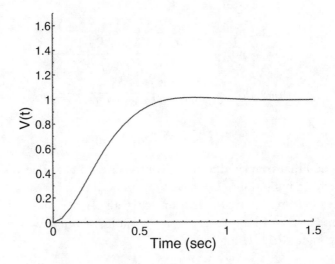

Figure 9.11: A step response of the transfer function in Equation 9.23 with proportional-integral control added.

The addition of integral control eliminates the steady-state error but the rise time was increased compared to the PD controller.

Lastly we will consider the full PID controller ($K_i = 300$ and $K_d = 50$ and $K_p = 350$)

$$G_{control}(s) = \frac{G_p(s)[K_p + \frac{K_i}{s} + K_d s]}{1 + H(s)G_p(s)[K_p + \frac{K_i}{s} + K_d s]} \tag{9.33}$$

$$G_{control}(s) = \frac{K_d s^2 + K_p s + K_i}{s^3 + (10 + K_d)s^2 + (20 + K_p)s + K_i} \tag{9.34}$$

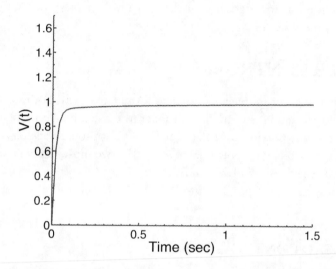

Figure 9.12: A step response of the transfer function in Equation 9.23 with proportional-integral-derivative control added.

In Figure 9.12 we have no overshoot, a fast rise time and no steady-state error. Remember, the goal of the controller is for the output to approximate the step function as closely as possible. For this example, the constants were specifically chosen to demonstrate a nicely tuned PID control system. In reality, many iterations were needed to find values that eliminated the steady-state error and overshoot but retain a fast rise time.

9.6   THE PROBLEM OF SYSTEM DELAYS

Hidden deep inside the PID controller is a serious problem. The controller is basing what to send the plant system on the *current* error signal. At each point in time it will compute a new m signal to send in and then use the feedback system to determine the effect. What happens if the plant system, $G_p(s)$, has a significant delay? Let us assume a one-second delay. Now assume that on the first time step, the controller sees a big error and so makes a large nudge. G_c then checks back in 0.1 second. But because of the delay, there has been no significant change in the error. So the next time (0.2 seconds), G_c makes an even bigger nudge. And this also will not register. In fact, G_c will continue to make the nudge larger and larger. In general, if the plant has a significant delay, the controller will continue to make m larger and larger until at some point an effect is seen. In our example, we will make 9 or 10 nudges before we even begin to see the impact of the first nudge! The result will be an enormous overshoot of the controller, possibly breaking the plant system.

The simplest method to overcome the effect of a plant delay is to slow down the rate at which the controller computes and sends the m signal. Specifically, we would want to update the controller *slower* than the delay in the plant. This way, we can be sure that m has had time to make its impact. A practical way to determine the delay in a real system is to send in an impulse to G_p and measure when the system first begins to respond.

9.7 OTHER CONTROLLERS

There are a number of other more exotic ways to formulate $G_c(s)$. For example, it is possible to add higher-order derivate terms to the D controller to make more accurate predictions of where the error will be. It is also possible to replace the P term with some non-linear term (like a cubic), in which case $e(t)$ is no longer proportional to $m(t)$. The two most popular alternative controllers, however, are *Fuzzy Controllers* and *Neural Network Controllers*. There are sometimes advantages of using these more complex types of controllers, but PID controllers still dominate the market due to their cost and utility.

9.7.1 LAG-LEAD CONTROLLERS

A cousin of PID controllers is the Lag-Lead controller. From the stand point of a bode plot, the PD controller is highpass, while the PI controller is lowpass. That means that the PID controller is either bandpass or band-attenuating (we will discuss these more in Chapter 12). But along with these changes in magnitude (top part of a Bode plot) come changes in phase (bottom part of Bode plot). Here the PD controller introduces positive phase (sometimes called *phase-lead*, while the PI controller introduces negative phase (sometimes called *phase-lag*). We will not go into the details, but Lag-Lead controllers are designed to add pairs of poles and zeros to "fix" the time properties of a PID controller.

9.8 REVERSE ENGINEERING BIOLOGICAL SYSTEMS

In Chapter 7 we discussed the idea that it is often helpful to be able to *open the loop* in a biological system. Here we are trying to remove some natural feedback in the system so that we can study the open loop system. Our example was Hodgkin and Huxley's Nobel Prize-winning work on reverse engineering how neurons generate action potentials. The problem was that the membrane voltage would change the level to which ion channels were open, which changed the current flowing, which then changed the membrane voltage. So we have the feedback loop that is shown in the top of Figure 9.13.

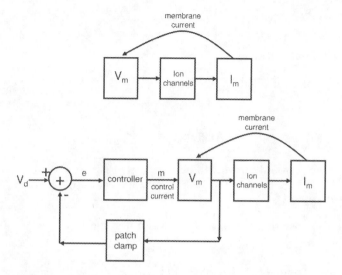

Figure 9.13: A greatly simplified closed loop of the Hodgkin-Huxley neural model on top. A simplified version of how the loop was opened by introducing a control current to counteract the membrane current.

In the bottom of Figure 9.13 is the addition of a control for the membrane potential V_m. The idea here was invented by Marmont and Cole, but then used by Hodgkin and Huxely to fix the value of V_m to anything they wanted. The idea was to first measure V_m—not simple to do in the 1940s and 1950s but it was possible using the patch clamp technique. The actual value of V_m could then be compared to the desired value, V_d, and an error signal generated. The controller would then create a control current (analogous to what we have been calling a nudge) that would be injected into the neuron. The idea here is that when $V_m = V_d$ we can perfectly balance the membrane current and control current flowing in the cell. In effect, they cancel one another out. A bit of thought will reveal that if the two are equal, and we know the control current, then we also know the membrane current. The experiment was complex, but in its most basic form, Hodgkin and Huxely were able to abruptly change V_d (like using a step input, called a voltage clamp) and then they could find the membrane current that resulted. What is most amazing about this is that the neuron is a very non-linear system with a lot of internal feedback, but their technique still worked just fine!

9.9   MATLAB

There are no explicit commands for feedback control within Matlab. Instead you can build up what you need using the *series, parallel* and *feedback* commands. A general strategy is to define your plant system first

```
>> num=1;
>> den = [1 10 20];
>> opentf=tf(num,den);
```

This is the system to be controlled

$$G_p(s) = \frac{1}{s^2 + 10s + 20} \qquad (9.35)$$

The step response can be found in Matlab by

```
>> step(opentf);
```

We see from the response that the open system is not going to cooperate so our next step is to put a feedback loop around it.

```
>> closetf=feedback(opentf,1);
>> step(closetf);
```

And again, we see that this did not allow the system to be controlled. Our next step would be to try a proportional controller.

```
>> Kp=300;
>> Ptf=tf([Kp],[1]);
>> Popentf=series(opentf,Ptf);
>> Pclosetf = feedback(Popentf,1);
>> t=0:0.01:4;
>> step(Pclosetf,t);
```

Here we have built up the system in a step-wise manner. Again, we see from the step response that we might do better by using a full PID controller.

```
>> Kp=350;
>> Ki=300;
>> Kd=50;
>> PIDnum=[Kd Kp Ki];
>> PIDden=[1 0];
>> PIDtf=tf(PIDnum,PIDden);
>> PIDopentf=series(opentf,PIDtf);
>> PIDclosetf=feedback(PIDopentf,1);
>> step(PIDclosetf);
```

Note that we used a little trick when we added the integral term—we needed to be sure that our transfer function has a $\frac{1}{s}$ term.

9.10 EXERCISES

1. What are some examples in your dorm room of devices that have a controller built into them? What type of controller do you think they are?

2. Arterial blood pressure, p(t), is regulated by a negative feedback loop involving barorecep-tors. A simple model of this regulatory system is Figure 9.14. Arterial blood pressure is mon-itored by baroreceptors whose output is compared to a reference pressure, $r(t)$. Brainstem neural mechanisms modulate blood pressure through control of sympathetic and parasym-pathetic activities to the heart and vasculature.

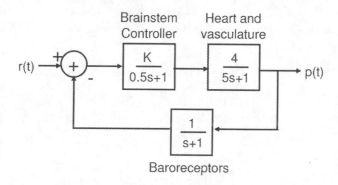

Figure 9.14: A very simple model of control of the vasculature.

    a. Analytically determine the open-loop transfer function relating $p(s)$ to the reference input, $r(s)$.

    b. Analytically, determine the close-loop transfer function relating $p(s)$ to $r(s)$.

    c. What is the differential equation relating the closed-loop $r(s)$ to $p(s)$?

    d. The brainstem controller includes an unspecified gain, K. If K is such that the trans-fer function of the close-loop system has poles on the imaginary axis, then $p(t)$ will oscillate even if $r(t)$ is constant. Find the value of K that will lead to an oscillation in blood pressure.

3. A two-compartment model of renal clearance was developed by Estelberger and Popper (2002) and is shown in the top of Figure 9.15. If something is wrong with the renal system, we can place I(t) under the control of an external device. This device will detect the con-centration in the central compartment (c) and adjust the input, I(t), to the renal system (as shown in the bottom of Figure 9.15).

    a. Derive the differential equations for the Estelberger and Popper without control. The variables are c and p.

b. Take the Laplace Transform of the two differential equations you derived in part a. Solve for the transfer function, $G_p(s) = \frac{c(s)}{I(s)}$ leaving the constants (k01, k12 and k21 as variables). It will be helpful to put this equation into the standard second-order form (i.e., the s^2 term has a coefficient of 1). You may assume that all initial conditions are zero.

c. Assume the following constants were found experimentally: $k_{01} = 0.0041$, $k_{12} = 0.0585$ and $k_{21} = 0.0498$. Show that the open-loop system, with the given values of the constants, is stable by showing the pole locations and the step response.

d. Create a P controller $(G_c(s) = K_p)$ that will stabilize the system at a concentration $c(t)$ equal to 100. A hint is that you will need to have $r(t) = 100$ and use the *lsim* command in Matlab. Note that the feedback gain, $H(s) = 1$. Demonstrate that your system is stable.

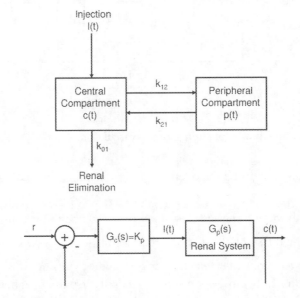

Figure 9.15: The two compartment model of renal clearance of Estelberger and Popper (2002).

4. You are designing a system that will use UV light to purify a population of cells. The idea is that you will send in some broad spectrum intensity of light, $L(t)$, that will then be converted, by $LF(s)$, into a UV source, as shown in Figure 9.16. From a previous experiment, you have determined that the transfer function is

$$LF(s) = \frac{0.5s + 4}{s + 1}$$

This UV light source will be sent to a dish of cells composed of a mix of Cell Type 1 and Cell Type 2. At the right intensity of UV light, Cell type 1 will survive while Cell type 2 will die. In a separate experiment you have determined that the relationships between UV light intensity and cell survival are

$$C_1(s) = \frac{s+3}{s^2+1}$$
$$C_2(s) = \frac{s+1}{s^2+1}$$

The three-block system, $G_p(s)$, will be your open loop plant systems. The output of this system is some level of a pure population, $p(t)$, of Type 1 cells.

a. Derive the Plant Transfer Function, $G_p(s)$, in the s-domain that relates the intensity of light to the purity of the population. Show your work.

b. Use Matlab to plot the open-loop poles and zeros.

c. Is the open-loop system stable, marginally stable or unstable? Explain.

d. If the light intensity is put under the control of a negative feedback system (with $H(s) = 1$), we can obtain a pure population of cells or $p(t) = 1$. The setpoint (goal) in this case will also be 1 (or a simple step function). Design a PD control system, $G_c(s)$, that will minimize rise time, overshoot and steady-state error. Report your values for P and D. Show the step response and 1) label the rise time, overshoot, 2% settling time and steady-state error on your plot and 2) report the values for rise time, overshoot, 2% settling time and steady-state error.

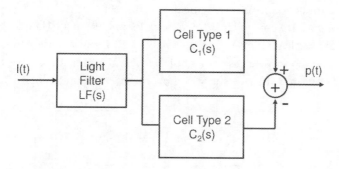

Figure 9.16: A simple model of the optical sorting of cells.

5. A social worker and epidemiologist are collaborating to develop a model that describes the impact of a mother and father's drug habits on the potential for their child to develop drug

habits as a teenager. They have collected some preliminary data on the individual impacts of the mother and father's drug use, as well as in combination. They have asked you to develop a model of this system and you sketch out the system in Figure 9.17. Based upon data you obtained from the epidemiologist you have derived

$$M(s) = \frac{s + 2}{s^2 + 1}$$

$$F(s) = \frac{s + 2}{s^2 + 10s - 3}$$

The input, $c(t)$, is any outside experiences (e.g., TV, music, friends) that may provide a sustained input (step function). The output, $d(t)$, is the state of the child. A stable child state is a teenager who does not use drugs, $d(t) = 1$. A marginally stable child state is one where the child is an occasional drug user, $d(t)$ oscillates or does not converge on 1. An unstable child state is one where the child becomes addicted to drugs (i.e., $d(t) \to \infty$ or $d(t) \to 0$).

a. If only the mother is raising the child will the child become an addict? Plot the step response.

b. If only the father is raising the child (i.e., $c(t)$ is directly input to $F(s)$), will the child become an addict? Plot the step response.

c. If both the mother and father are present, will the child become an addict? Plot the step response.

d. The social worker would like to predict the potential impact of therapy by a psychologist. There are many different therapy styles (tough love, overwhelming the child with positive influences). As an engineer, you suspect that the psychologist is similar to introducing a PID controller in a feedback loop. Design a PID controller that will result in a more stable system. You must give your P, I and D parameters (i.e., therapy technique) and report your assessment of the system response (steady-state error, rise-time, 2% settling time and overshoot). Show these values on a plot of the step function.

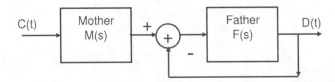

Figure 9.17: A hypothetical model of the interactions within a family unit.

CHAPTER 10

Time Domain Analysis

Signal Processing is a term used by engineers to describe taking a signal and performing some function on it. This could be for any one of a number of reasons but almost always the intent is to extract some particular information. For example, we may be confronted with a time series that represents the electrocardiogram from a patient and we would like to know their heart rate. Doing so will involve finding the QRS complex and marking the peak. Then to get the heart rate, we need to be able to measure the time delay between peaks.

In a more abstract example, let's assume that we have a signal, $x(t)$, which was output from a complicated system with many internal sub-systems that all contributed to the signal.

$$x(t) = a(t) + b(t) + c(t) + \ldots + z(t) \tag{10.1}$$

Here the signals $a(t)$ through $z(t)$ are components of the recorded signal $x(t)$. The problem is that we don't have all of the components of $x(t)$. What if we wanted to extract component $q(t)$? In general, we can't extract $q(t)$ unless we know something about what it might look like, or what all of the other terms are. Luckily, in most biomedical situations, we do have some idea of what we are looking for.

To make this more concrete, let's assume our biomedical signal, $x(t)$, is the oxygenation level from pulse oximetry. What we want is a measure of the oxygen levels in the body, but the signal we record has combined with it a pulse, noise from skin, bone, maybe fingernail polish, muscle motion of the finger, electrical noise from the battery, electrical noise in the room, as well as some of the electrocardiogram and respiration signals. In reality, many of these signals will contribute very weakly, but together they will obscure what we want—the level of oxygenation.

In this chapter, we will explore some common ways to process signals to extract information. Something important to keep in mind is that the overall goal in analyzing a signal is to learn something about the nature of the system. For example, we may want to know something about delays, or how much memory it has. We might even want to estimate something about how stable the system is. Later in the chapter, we will explore *correlations* between signals to determine if they are related, and if so, how strongly. Again, in finding correlations, we are really after information regarding how related the sources of the signals are to one another within the system. For example, it has been found that some forms of epileptic seizures are preceded by irregular heart beats. This information was found by correlating EEG from the brain and ECG from the heart. Such correlations could be useful in designing devices to predict and manage epilepsy.

10.1   BASIC SIGNAL PROCESSING

There are some methods that you use already that can be considered signal processing. In general a time-domain signal is a vector of numbers, with each number representing the signal at a particular time. It is from this vector we wish to extract information. The format of that information might be a single number, some other vector, or in more advanced signal processing some other format (e.g., image, two vectors, tensor). Nearly all signal processing methods can be thought of as being applied to continuous or discrete signals. In this chapter we will focus primarily on discrete signals because basic signal processing concepts are often easier to understand in the discrete Time Domain.

10.1.1   AVERAGE

The *time average* of a signal is certainly the most simple and widely used type of processing. By our most broad definition (extracting information) the average condenses a large amount of data down to a single number. First, we will define our signal vector as

$$x[n] = [x_1, x_2 \ldots x_{N-1}, x_N] \tag{10.2}$$

Note that to make things easier to compute, we have transformed $x(t)$ (a continuous time signal) into $x[n]$ (a discrete time signal).

The average of $x[n]$ is then defined as:

$$\bar{x} = \frac{1}{N} \sum_{i=1}^{N} x_i \tag{10.3}$$

The signals and systems term that is often used in place of average is the *baseline* or *DC offset*. Figure 10.1 shows two data sets from the same individual. The top trace is the heart rate over time while resting. The bottom plot is the heart rate while meditating. Visually we can see that they are different. But our first important measure is to take the average. The average for the resting rate is 66.5 beats/minute while the meditating rate is 81.3 beats/minute. It is somewhat surprising that the meditating rate is so much higher!

Note that here we have taken a large string of numbers (a vector) and reduced all of the information down to a single number (scalar). As such, we have lost a great deal of information. For example, if you received only an average, there would be no way to reconstruct the original signal. There are many types of processing where information is lost. In fact, we can think of signal processing as passing a vector as an input to a block diagram (in this case a signal vector to a block that performs the average). Some processing blocks perform invertible operations, while others do not.

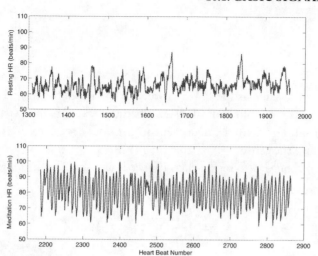

Figure 10.1: Time series plots for two data sets. The top data set is the normal resting heart rate (beats/minute) for a resting individual. The bottom plot is data from the same individual but while meditating.

10.1.2 SIGNAL POWER

It is often assumed that any deviation from a baseline requires energy from the system. For example, if a circuit pushes a voltage up and down in time, somewhere it is using a source of energy to do so. The *signal power* is a measure of deviation from baseline and is defined as

$$X^{RMS} = \sqrt{\frac{1}{N}\sum_{i=1}^{N}x_i^2} \tag{10.4}$$

Again, the assumption here is that the system must expend energy to cause the signal to deviate from the 0 line. Another name for the signal power is the *Root Mean Square* or RMS power. Again note that we have taken a sequence of points (a vector), and reduced them down to a single number (a scalar). In the process we have lost some of the information in the signal.

10.1.3 VARIANCE AND STANDARD DEVIATION

The problem with RMS power is that it is measured from a baseline of zero. Consider a signal that doesn't vary at all but has DC offset. If you apply Equation 10.4, you will have a non-zero RMS power and it will appear that your system expended a great deal of energy to generate the signal. In reality the system was most likely stable at the DC offset and would expend no energy to stay there. To compensate for an offset we can define the *variance* from the average.

$$\sigma^2 = \frac{1}{N} \sum_{i=1}^{N} (x_i - \bar{x})^2 \tag{10.5}$$

Here the variance is the same as the RMS power but the mean (offset) has been subtracted out. This idea is very important in biomedical systems because there are many parts of devices that can contribute to an offset. For example, many sensors introduce some form of offset at the interface between the body and the sensor.

The more common way to report the variance is as a *standard deviation* which is defined as

$$\sigma = \sqrt{\frac{1}{N} \sum_{i=1}^{N} (x_i - \bar{x})^2} \tag{10.6}$$

Returning to Figure 10.1, the standard deviations for the resting and meditating data are 5.3 and 9.3 respectively. Again, notice that we have condensed a large amount of data down to a single number. We also have a more physical interpretation (as opposed to an abstract statistical interpretation) of the meaning of standard deviation and variance as related to signal power and energy within the system. The meditation ECG has more variance.

10.1.4 SIGNAL TO NOISE RATIO

In the introduction we discussed the idea that a recorded signal, $x(t)$, may be thought of as the sum of several other signals.

$$x(t) = q(t) + a(t) + b(t) + \ldots \tag{10.7}$$

Here we have separated our signal into two parts, the desired part, $q(t)$, and everything else. We can represent this idea for any signal as

$$x(t) = x_{desired}(t) + x_{noise}(t) \tag{10.8}$$

There is an important point here. The signal we do not want is generally called *noise*. But the meaning of noise is in the eye of the engineer. If we were after the signal $m(t)$, then $q(t)$ would be considered part of the noise! In Chapter 12, we will discuss different methods of separating signal from noise. For now we are after some measure for the relative magnitudes of the signal and noise. *Signal to Noise Ratio*, or SNR, is a ratio defined by

$$SNR = 20 \log\left(\frac{X_{signal}^{RMS}}{X_{noise}^{RMS}}\right) \tag{10.9}$$

Where RMS is the Root Mean Squared power of the signal as defined by Equation 10.9. We can see that if the signal is large compared to the noise, then the SNR is also large. Likewise as the level of noise grows, SNR becomes smaller. In other words, higher SNR means the signal is easier to pick out from the noise.

10.2 CORRELATIONS

So far, we have considered operations on one signal. But a common question is how similar two (or more) signals are to one another. For example, in recording an EEG we might collect 36 signals from the scalp. So we might ask which of these signals have something in common. After all we know they are outputs from the same brain. Here, again, we are trying to probe into the system to find out which parts of the brain are functionally connected together, or working in collaboration with other areas of the brain. In other words, we want to know how well *correlated* these areas are.

If we have two signals, $x(n)$ and $y(n)$, the ideal would be to have a single number to tell us the *cross-correlation*. In reality there are all sorts of ways they might be correlated, and some of them would involve all sorts of complex signal processing in addition to several numbers. A good first place to start is a very simple type of correlation.

$$C_{xy} = \frac{1}{N} \sum_{i=1}^{N} x[i] y[i]$$ (10.10)

(10.11)

Here we are simply lining up the signals and then multiplying each point together (Figure 10.2). We can consider three different cases. First, let's assume the two signals are exactly the same. In this case the correlation will be a large positive number. Second if the signals are exactly the opposite, we get a large negative number. Another name for this case would be that it would be *anti-correlated*. Lastly, if the signals are dissimilar then we would get a sum near zero. The reason is that the places where they line up and where they don't line up would be entirely random and any correlations would be canceled out by anticorrelations.

There are some problems with this measure that we will now correct in a series of steps. First, what happens if $x(t)$ is much bigger in magnitude than $y(t)$? In this case any correlation could become lost in the differences in amplitude. Another way to see this is to ask: what happens when we get a number, say 34.65098, from the calculation? Is that good correlation or not? Is it close to zero or far away?

What we are really after is if the *shapes* of $x(t)$ and $y(t)$ are similar or not. So to refine our correlation measure, we can first *normalize* the signals.

$$C =$$

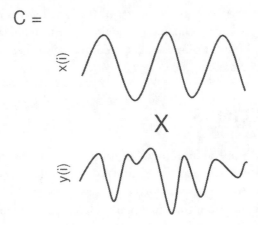

Figure 10.2: Correlation of two signals using Equation 10.10.

$$C^{norm} = \frac{C}{\sqrt{\sigma_x^2 \sigma_y^2}} \qquad (10.12)$$

where the σ^2 terms are the variances of the two signals. Here the maximum correlation is 1 when the signals are identical, -1 when they are exactly the opposite (anti-correlated) and 0 when the are not correlated.

10.2.1 CROSS-CORRELATION

A different problem occurs when we consider the correlation between x as a sine wave and y as a cosine with the same frequency. A bit of thought will reveal that you are going to get a summed value of zero! It may have seemed strange that two sinusoids (sin and cos) with the same frequency would turn out to be uncorrelated. Our instincts would seem to tell us that the sin and cos functions have the same shape and so should be related in some very strong way. The problem is that they are the same shape but shifted by 180 degrees. In a biomedical setting, this type of shifted correlation could easily occur in two signals, such as the EEG, where the signal at one site follows the signal at another site.

It is important to pause before we get into the math and reflect on the idea in the paragraph above. What we are really talking about is the idea of delays in a system. If we see that some feature shows up in a signal $x(t)$ and then that same (or similar) feature shows up at some later time in $y(t)$, what does that say? First, it says that they *seem* to be related. But we might now be able to measure the delay between them. This gives us some information about the system. For example, if our two signals are from the brain, it might tell us how long it takes neural impulses to travel from one site to the other. If the original locations are close together it may even tell us something about how much processing has taken place (longer delays would allow for more processing to occur).

Now there is one very important trap that is waiting, and has snared many otherwise very smart scientists. *Correlation Is NOT Causation.* Just because $x(t)$ preceded $y(t)$ does not mean that $x(t)$ *caused* $y(t)$. It may be that there is some third signal, $z(t)$, that is generated elsewhere in the brain that is the root cause of both signals. In other words, it may be that the brain areas generating $x(t)$ and $y(t)$ are not connected at all! Look out for this type of trap in both science and engineering. It is often passed off as a logical argument, but you cannot claim causation just by showing correlation.

We can now return to the problem of phase shifts in our sinusoids. To do so, we will define the *cross correlation* as

$$C_{xy}(k) = \frac{1}{N} \sum_{i=1}^{N} x(i)y(i+k) \tag{10.13}$$

where k is a delay variable. Here we are performing a series of correlations (as in Equation 10.10) but for many different delays (values of k). You can think of this as keeping $x(t)$ the same and then sliding $y(t)$. After each slide, you compute a correlation. Then slide again and compute another correlation (see Figure 10.3). In this way, we can create a plot of $C_{xy}(k)$ versus k, as in Figure 10.4.

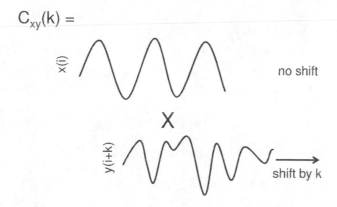

Figure 10.3: Building the cross-correlation of two signals using Equation 10.13. Only shown is building one value of $C_{xy}(k)$ for the delay k.

The important point on the plot in Figure 10.4 is the maximum. Here the maximum shows the best correlation we can possibly get. And we have the added benefit that we even can get a good estimate of the delay between the two signals if the correlation is good.

We can now return to the idea of the sin and cos functions. Imagine holding the sinewave constant and shifting the cosine; eventually they will line up and give a cross-correlation of 1. But

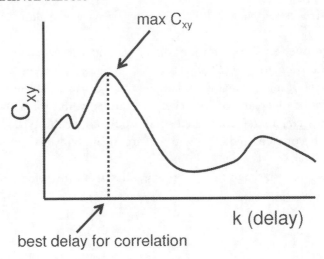

Figure 10.4: Cross-correlation as a function of delay. Each point in the plot is created using the method in Figure 10.3.

we will be able to see that this was achieved only for a shifted version of the signal (when C_{xy} reaches a maximum).

10.2.2 CROSS COVARIANCE

A bit of thought will show you that just as the intent of signal power (Equation 10.5) can be ruined by an offset, the correlations we defined are also susceptible to the same problem. We therefore must account for differences in the averages in our signals by subtracting them out.

$$CV_{xx}[k] = \frac{1}{N} \sum_{i=1}^{N} [x[i] - \bar{x}][x[i + k] - \bar{x}] \qquad (10.14)$$

$$CV_{xy}[k] = \frac{1}{N} \sum_{i=1}^{N} [x[i] - \bar{x}][y[i + k] - \bar{y}] \qquad (10.15)$$

A second problem that can occur in a real situation is if two different measuring devices were used (e.g., different EEG electrodes). In this case, one signal may be much larger than another. Large differences in amplitude can also obscure an accurate correlation measure. To prevent this problem it is often useful to *normalize* signals first.

10.2.3 AUTO CORRELATION

Cross-correlations are a simple means of detecting similarities between two signals. But what if we turned this idea around and shifted a signal with respect to itself? In other words, what if we performed a cross-correlation but in place of $y[n]$, we just used a copy of $x[n]$?

$$C_{xx}[k] = \frac{1}{N} \sum_{i=1}^{N} x[i]x[i+k] \tag{10.16}$$

If this plot is normalized, the signals will be perfectly correlated for a delay of 0 and therefore $C_{xx}[0] = 1$ by definition—a signal is perfectly correlated with itself when there is no delay. But now consider what happens when we slide the signal against itself. Another way to think of this is that we are checking to see how well correlated a signal is with later versions of itself.

As a first intuitive step, imagine performing an auto correlation on a sinewave. It will be perfectly correlated when the delay is 0. However, as we slide y, the signals become less and less correlated until they are not correlated at all (as in our example above for sin and cos functions). But, if we continue to slide, they become correlated again. Then less correlated. And so on.

But there is something interesting embedded in this idea of seeing how well correlated a signal is with a later version of itself. If we think of the time series for a signal as

$$[x_0, x_1, x_2, x_3, x_4 \ldots] \tag{10.17}$$

then what we are really doing is finding out how well correlated, say, x_{27} is with x_3. In most systems, we can imagine that two values that are next to one another are probably correlated pretty well. So x_{27} is probably correlated with x_{26}. The reason is because often real systems (remember they are the source of these signals) have *memory*. What this means is that in some way the state of the system that generated x_{26} is still reverberating throughout the system when x_{27} is generated. So an important question is how long after the system generates some signal is it still influencing future signals? This is just another way of saying: how long is the memory of the system?

In fact, the auto correlation gives you an estimate of the memory of a system. When the delay is zero, of course $C_{XX} = 1$. And we would expect that the signal is well correlated to the point that came before. But a signal at a particular time will be dependent on more than just what happened at the time before. It will be correlated with some number of points in the past. And that is exactly what the auto correlation is checking—how far into the past there is a correlation.

Now, in many systems (even ones that do not oscillate), there will be repeating patterns. The problem is that the reoccurrence of these patterns can fool the auto correlation, making correlations appear for very long delays, when in fact the memory of the system is clearly not that long. For many signals, however, there is a point where the correlation crosses from positive to negative. It is the delay that corresponds to this *first zero crossing* that is often used as an estimate of the memory of the system. A little thought will show that when the auto correlation plot crosses 0 for the first time, we have the copy of x far enough away that it is no longer correlated to x.

A real example is shown in the auto correlation of the resting and meditating heart rate shown in Figure 10.5. The meditative plot oscillates because it repeatedly is correlated and then anti-correlated for different delays. We can somewhat even pick this up by looking at the time plot, because it nearly oscillates. But the first zero crossing for meditation occurs for a small delay, implying that it has very little memory. The normal heart rate, on the other hand, seems to have a very long memory—it remains correlated for a long time.

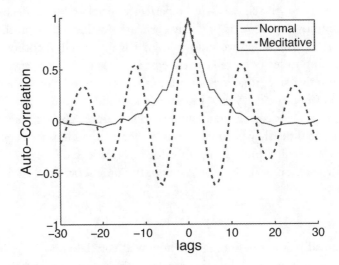

Figure 10.5: The auto-correlation of the data in Figure 10.1.

One last word must be mentioned about memory. You may have noticed that the word *estimate* was used. The reason is because when we are looking at an auto correlation, we are only looking at one signal from the system. And that one signal gives us a value for memory. But if we looked at another signal, it might give us a different value. So another way to think of the memory *estimate* is as a lower bound—we estimate the system has at least this much memory.

10.3   MATLAB

Matlab has an enormous number of built in functions that will allow for Time Domain processing. We will only cover the most basic functions that correspond to the topics covered in this chapter.

We will assume that we have two data vectors, x and y, and then list the possible operations that might be performed. Note that some functions will take one vector while others (usually the ones that find correlations) will take two signals. See the help file for the functions to learn more about the various options.

```
>> x = [0 4 -3 5 8 27 1 24];
>> y = [1 5 2 3 6 0 12 32];
>> mean(x); % mean of x
>> var(y); % variance of y
>> mode(x); %mode of x
>> median(y); %median of y
>> std(x); %standard deviation of x
```

There are some functions, such as *xcorr* and *xcov*, in Matlab that will take cross and auto correlations, but they have many options and it is sometimes unclear how they work. Instead a file, axcor.m, is provided (on the website) that will allow you to take cross correlations and autocorrelations.

Let us assume that the two heart rate datasets shown in Figure 10.1 were stored in the two variables *x* (Meditative) and *y* (Normal Resting). Then we could use *axcor.m* as follows to generate Figure 10.5.

```
>> [pre,prelags]=axcor(y-mean(y));
>> [med,medlags]=axcor(x-mean(x));
>> h = plot(prelags,pre,'k');
>> hold on
>> h = plot(medlags,med,'--k');
```

Note that we have subtracted out the mean before taking the autocorrelation. When only one argument is passed to *axcor* it assumes to take the autocorrelation of the signal. The output arguments are the correlation data and the shift (sometimes called the *lag*). We can then plot to the correlation and lag against one another to generate the autocorrelation plot.

If instead we wanted to find a cross-correlation with some other signal (e.g., a sinusoid), we could again use the *axcor* function but with two input arguments.

```
>> f = 30;
>> t = (1:length(x))./length(x);
>> sinusoid1 = sin(2*pi*f.*t); %Create a sinusoid
>> x_norm = x-mean(x);   %Subtract DC Offset
>> MinSig = min(x_norm);   %Compute Minimum
>> MaxSig = max(x_norm);   %Compute Maximum
>> x_norm = 2*(x_norm-MinSig)/(MaxSig-MinSig)-1;  %Normalized Signal
>> r = axcor(x_norm,sinusoid1);   %Compute correlation vector
>> max(r)   %Find the maximum correlation
```

10.4 EXERCISES

1. What are two examples from your life where you would expect two signals to be strongly correlated? What is one set of signals from your life that are strongly anti-correlated?

2. An MD calls you on the phone and asks for your help with the following problem. She has recorded an electromyogram (EMG) using a new device. Unfortunately, not all of the displays are working. The two numbers she knows are the RMS value of the noise ($0.3V$) and the Signal-to-Noise (10dB). Perform a calculation to give the MD the RMS value of the EMG. Give your answer in volts.

3. Derive the correlation (unnormalized and unshifted) between $\sin(2\pi t)$ and $\cos(2\pi t)$ over some interval 0 to T. A well-chosen trig identity may make the derivation easier.

4. Given an auto-covariance plot that is always 1, explain what the signal looks like.

5. Given the two auto-covariance plots in Figure 10.6, explain

    a. Which signal is more periodic? Why?

    b. Which signal has the most memory? Why?

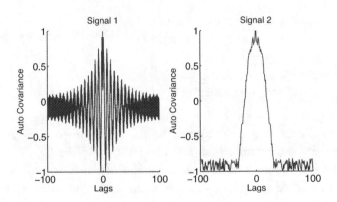

Figure 10.6: Two example autocorrelations.

6. In 2000, Goldberger *et al.* published a study comparing heart rate variability from resting and meditating individuals. These data were posted on http://www.physionet.org, a repository for example data sets collected from the body. You can download these datasets (*HR_med.mat* and *HR_pre.mat*) from the website.

    a. Use the Matlab "load" command to load these files into memory. You should see four arrays that represent time and the signal. Plot t_pre versus hr_pre (normal) and t_med versus hr_med (meditation) on separate figures that are the same as shown in Figure 10.1. Visually you can see a difference but we want to quantify this difference.

    b. Use Matlab to report the average heart rate, standard deviation and variance for the normal and meditative states. Present your solutions in a table. You can check your answers by looking them up in the chapter.

c. You next must remake Figure 10.5 to assess differences in the memory of the normal and meditative states. First, clear the Matlab memory and reload HR_med.mat and HR_pre.mat. Then download the file *axcor.m*. As explained in the text, *axcor.m* will compute both auto and cross correlation. Following the example above, subtract out the mean for each signal. Then make a plot of the time delay (lag) versus the auto covariance (because you have subtracted out the mean) for the normal heart rate. Remember that this plot should be symmetric about Lag=0. You should show lags between −30 and 30 (using the *axis* command). Do the same for the meditative heart rate and present both sets of data on the *same* figure (use the *hold on* command in Matlab).

d. Make a plot of the cross-correlation (again use *axcor.m*) between the normal and meditative data and sinewaves that vary in frequency from 1 to 200Hz. Your plot should contain the frequency on the x-axis and correlation with the Heart Rate on the y-axis. Hint: You can loop through several sinusoids with code that has the following form:

```
Normalize the signal y
time = 1:dt:endtime
for f=1:200
        y = sin(2*pi*f.*t);
        COMPARE y and your signal with cross correlation
    Find maximum of the cross correlation (use Matlab ``max'' command)
        Save the maximum into an array (e.g., rmax(f) )
end
plot(f,rmax(f)
```

Please turn in both your plot and your Matlab code.

CHAPTER 11

Frequency Domain Analysis

In the previous chapter the idea of cross-correlation was developed to determine how similar two signals are to one another. And in Figures 10.2 and 10.3 we used a sinewave as one of the signals. It turns out that this is a very fruitful and important way to think about decomposing a signal—by comparing a complex signal to many different sinewaves (of different frequencies) we can identify the frequency content of the complex signal. This is particularly important in a biological context where many of the signals are complex and where certain diagnostic procedures rely on measuring changes in the frequency contributions. In this chapter we will show how making this type of comparison leads to some very powerful mathematical analysis techniques in the frequency domain.

11.1   COMPARING A SIGNAL TO SINUSOIDS

We will first consider comparing a somewhat complex signal to sinusoids of different frequencies.

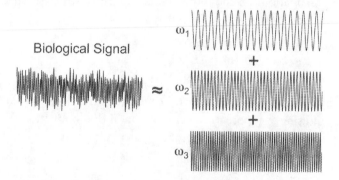

Figure 11.1: Representing a complex biological signal as the sum of sinusoids.

A natural question is which frequency sinusoid is correlated best. Doing so would tell us how to approximate our signal with a single sinewave. Visually, it is clear that there is no one sinewave that will make a perfect approximation. But what if we use the two sinusoids that are best correlated? Or three? Or more? Here we could simply add up the sinusoids and scale their amplitudes by the degree to which they are correlated. So for the two sinusoids

$$x_{approx}(t) = a_2 \sin(\omega_2 t) + a_1 \sin(\omega_1 t) \tag{11.1}$$

where ω_1 and ω_2 are the two best correlated frequencies. The constants a_1 and a_2 will be in some way proportional to the degree of correlation. Again, you might imagine that two sinusoids will not make a very good approximation. However, consider what will happen if we used three or four or n sinusoids.

$$x_{approx}(t) = a_n \sin(\omega_n t) + a_{n-1} \sin(\omega_{n-1} t) + \ldots + a_2 \sin(\omega_2 t) + a_1 \sin(\omega_1 t) \qquad (11.2)$$

A little thought might convince you that, at least in principle, by adding more sinusoids to our approximation, we could recreate an arbitrarily complex signal. The opposite way to think about this is that we have *decomposed* our signal into a number of sinusoids. Mathematically the continuous correlation function is

$$C_{xy}(\tau) = \frac{1}{T} \int_0^T \sin(\omega t + \tau) x(t) dt \qquad (11.3)$$

Where $x(t)$ is the signal we wish to probe. After performing the cross-correlation, we have a plot of C_{xy} versus τ. The peak of this plot will tell us the best correlation that we might have which we may call C_{xy}^{max}. However, we still have the variable ω in Equation 11.3. By varying, ω we can make a plot C_{xy}^{max} versus ω, or $C_{xy}^{max}(\tau)$.

As a first approximation, we could then define our reconstructed signal as:

$$x_{approx}(t) = \sum_{n=0}^{N} C_{xy}^{max}(\omega_n) \sin(\omega_n t) \qquad (11.4)$$

where N is the number of frequencies (ω) we have tested. Notice that Equation 11.4 is very similar to Equation 11.2 but using the more compact notation of the summation. Very often we will try a number of frequencies, starting at zero, ending as some high frequency, f_h, with intervals in between, Δf. The equation then becomes

$$x_{approx}(t) = \sum_{n=0}^{N} C_{xy}^{max}(\omega_n) \sin(n \Delta 2\pi f_o t) \qquad (11.5)$$

Where f_o is some base frequency and $n \Delta 2\pi f_o$ is a multiple of that fundamental frequency.

11.1.1 PROPERTIES OF SINUSOIDS

You may be wondering why we would be so interested in comparing a signal to sinusoids. The reason is that sinusoids have some very unique properties. The first property is that sinusoids of different frequencies are all independent (i.e., orthogonal) of one another. This is the reason why we can approximate a complicated signal by adding up more and more sinusoids. In fact, the more sinusoids added, the better the approximation will be. Second, a pure sinusoid has energy at only one frequency. This is very much related to the idea that sinusoids are independent of one another. Third, the physical meaning of a sinusoid is easily understood by humans. Consider that colors are simply different frequencies of light and that tones are simply particular frequencies of pressure waves in the air. Lastly, when a sinusoid is input to a linear system, the output will be a sinusoid of the same frequency (see Chapter 8). And we know how to characterize the response with the Bode plot. For all these reasons, decomposing a complex signal into sinusoids is the path we will follow in this chapter.

11.1.2 A PROBLEM WITH THE CROSS-CORRELATION

So far we have built up the idea that we could simply compare a complex signal to a number of sinusoids, checking the cross-correlation of each one, to arrive at the amplitudes, a_n. In other words, we have been arguing that

$$C_{xy}^{max}(\omega_n) \approx a_n \qquad (11.6)$$

But there is problem with this idea that we can only get at by trying to apply it. Let's assume we have a known signal

$$f(t) = 1.5 \sin(100 \times 2\pi t) + 1.75 \sin(200 \times 2\pi t) + 2.0 \sin(300 \times 2\pi t) \qquad (11.7)$$

Now imagine that we find the correlation of $f(t)$ with frequencies between 10Hz and 500Hz in increments of 10Hz. For each frequency we will get a value of C_{xy} and we can then plot that against the frequency tested. The left panel of Figure 11.2 shows the result.

You will notice that although the amplitudes are not exact, we do get a spike (good correlation) at the proper frequencies, and we even get some relative measure of the amplitudes.

Now the problem is that in principle, we wouldn't know the origin of the signal before hand. What if

$$f(t) = 1.5 \sin(100 \times 2\pi t) + 1.75 \sin(204 \times 2\pi t) + 2.0 \sin(306 \times 2\pi t) \qquad (11.8)$$

In Equation 11.8 the last two sinusoids have a slightly higher frequency. If we apply the same procedure of cross-correlation, we arrive at the plot in the right side of Figure 11.2. And here is our problem. There are virtually no peaks at 204Hz and 306Hz. And the reason is that we didn't test those frequencies, we tested 200Hz and 300Hz. Now a possible solution is to change the increment from 10Hz to 1Hz. Then we would have tested 204Hz and 306Hz. But it would come at a cost—testing in increments of 10Hz required 50 tests. Testing in increments of 1Hz would require 500 tests. On a computer this could become very significant. Especially if our actual frequencies were 204.5Hz and 306.7Hz, requiring increments of 0.1Hz.

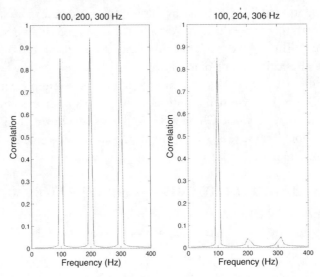

Figure 11.2: Cross-correlation in left panel for Equation 11.7 (left) and 11.8.

11.2   THE FOURIER SERIES

The problem encountered using the cross correlation is a concern because it shows a glimmer of usefulness, but doesn't exactly work. We are going to introduce the beginning of a fix in the Fourier Series and then make it more complete in the Fourier Transform.

It all started with Joseph Fourier who was studying the flow of heat down a metal cylinder. He invented an idea very much related to our idea of correlation of sinusoids, the Fourier Series, to help him interpret his results. He started by assuming that the unknown signal, $y(t)$, was periodic (i.e., repeated itself over and over again) with a period of T. Consider again the cross-correlation

$$C_{xy}(\tau) = \frac{1}{T} \int_0^T y(t)x(t+\tau)dt \tag{11.9}$$

We can then let $x(t+\tau) = \cos(\omega_o t + \theta)$, where ω_o is some *fundamental* frequency and θ is functioning like our delay (τ). The meaning of the fundamental frequency needs a little explanation. Let's assume that we have the signal $x(t)$ that repeats every $T = 5$ seconds. Then the fundamental sinusoid would have a frequency that would cover the entire signal in one period. So the fundamental frequency needs to go through one cycle in 5 seconds, and its frequency will then be

$$f_o = \frac{1}{T} \qquad (11.10)$$

$$\omega_o = \frac{2\pi}{T} \qquad (11.11)$$

Note that in Figure 11.3, ω_o completes one full cycle over the entire signal. We can then find the correlation, r, between this fundamental frequency (see Figure 11.3) and the signal as

$$r(\theta_o) = \frac{1}{T} \int_0^T y(t) \cos(\omega_o t + \theta_o) dt \qquad (11.12)$$

But we may not have lined up our fundamental frequency in the best way in Figure 11.3 so we can shift around θ_o to find the best correlation, r_{max} at the fundamental frequency.

Biological Signal

ω_o

$2\omega_o$

$3\omega_o$
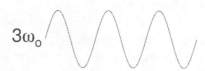

Figure 11.3: Relationship between a signal and the fundamental frequency ω_o and the first two harmonics.

The next step is to check the correlation with $2\omega_o$, then $3\omega_o$ and so on. In general we want to check $n\omega_o$ where n is some integer. This series of frequency multiples of the fundamental are known as *harmonics*.

$$r_{max}(\omega_n) = MAX\left[\frac{1}{T}\int_0^T y(t)\cos(n\omega_o t + \theta_o)dt\right] \tag{11.13}$$

So Equation 11.13 is simply testing for the best correlation (by sliding around θ) for a series of sinusoids, all of which are some multiple of a fundamental frequency, ω_o.

To demonstrate the idea, think about how to represent something like a step function. In Figure 11.4 when we use just three harmonic frequencies (ω_o, $2\omega_o$ and $3\omega_o$) we get a somewhat poor representation. But as we include more harmonics, we get a better and better representation.

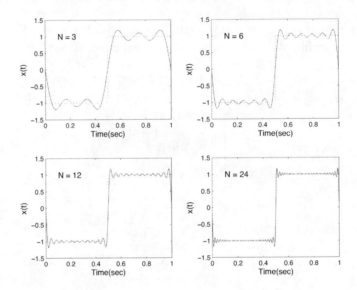

Figure 11.4: Using more and more harmonics to represent a step function.

You may notice that near the corners of the step we get some ringing. This is called *Gibb's Phenomenon* and is seen whenever we try to represent a signal that has sharp corners. The intuitive reason is that sharp corners require almost infinitely high frequencies to represent perfectly. So we would need to include infinitely many harmonics to get it perfect.

So the bottom left panel of Figure 11.4 shows what happens when we represent a step with 24 harmonic sinusoids. But the amplitude of those sinusoids are not all the same. We can get the amplitudes from Equation 11.13. And our approximated signal can now take the form of Equation 11.2.

$$\begin{aligned} S_{approx} &= r_{max}(\omega_o)\sin(\omega_o t) + r_{max}(\omega_2)\sin(2\omega_o t) + \\ &\quad \dots + r_{max}(\omega_{23})\sin(23\omega_o t) + r_{max}(\omega_{24})\sin(\omega_{24}t) \end{aligned} \tag{11.14}$$

Now we might ask how to encapsulate all of the information about the coefficients that correspond to particular frequencies. And in fact, the plots we will encounter will be exactly that—plots of the frequency on the x-axis with values that indicate the amplitudes on the y-axis.

11.3   THE FOURIER TRANSFORM

There was an assumption hidden inside the Fourier Series that just won't work for biological signals. They aren't periodic! In fact, few signals (other than mathematical signals) are. So what we need to do is fix up the Fourier Series so that it will work on *aperiodic* signals. In doing so, we will arrive at the Fourier Transform.

The trick is actually very simple—just assume an infinite period (i.e., $T \to \infty$). A consequence is

$$\frac{1}{T} \to 0 \tag{11.15}$$
$$\omega_o \to 0 \tag{11.16}$$

Now the harmonics are going to be $n\omega_o$. And it may seem that everything is zero, but remember we are talking about limits here. So at the same time $\omega_o \to 0$, $n \to \infty$. In the limit $n\omega_o$ becomes a continuous variable ω

$$lim_{T \to \infty} r_{max}(\omega_n) = \frac{1}{T} \int_0^T y(t) \cos(n\omega_o t + \theta_o) dt \tag{11.17}$$

$$r_{max}(\omega) = F(\omega) = \int_{-\infty}^{\infty} y(t) e^{-j\omega t} dt \tag{11.18}$$

Notice that the definition of the Fourier Transform is very similar to Equation 11.3 above where the sinusoid has now been expressed in exponential notation (see Appendix A).

And we can go one step further to define this operation on the signal $y(t)$ as a transform

$$F[y(t)] = \int_{-\infty}^{\infty} y(t) e^{-j\omega t} dt \tag{11.19}$$

You may have noticed that the definition of the Fourier Transform is almost exactly the same as the Laplace Transform—just let $s = j\omega$. In fact, the Fourier transform is sometimes used by engineers (and physicists/mathematicians) to solve differential equations in the same way as the Laplace Transform. In a similar way, a signal in the Time Domain is *mapped* to a new domain. Here the new domain is the *frequency domain* (i.e., ω-domain). Again, the reason for transforming from the Time Domain to the frequency domain is because we have a more intuitive feel for the frequency domain.

11.3.1 POWER AT A FREQUENCY

In general the Fourier Transform of a real signal will have both a real and an imaginary part. To plot the result we often use what is known as Parceval's Relationship and simply plot the *power* of the signal.

$$PS(f) = \left| X(f) \right|^2 \tag{11.20}$$

Where $PS(f)$ is known as the *Power Spectrum* of the signal, that will include some contribution from both the imaginary and real part of the solution. While this loses some information (especially about phase), it is much easier to understand and in most biological contexts is all we need.

11.3.2 FOURIER TRANSFORM PROPERTIES

The Fourier transform, like the Laplace Transform, is a linear operator. This means that if it is applied to a signal from a linear system, it will not destroy the property of linearity. For example,

$$F[y(t) + x(t)] = F[y(t)] + F[x(t)] \tag{11.21}$$
$$F[ay(t)] = aF[y(t)] \tag{11.22}$$

There are a number of other properties that you can see in Appendix D. Each can be derived by working through Equation 11.19. A few notable transforms are the time shift

$$F[y(t - \tau)] = e^{-j\tau\omega} F[y(t)] \tag{11.23}$$

and the step function, $u(t)$

$$F[u(t)] = \frac{1}{2} \left[\delta(\omega) - \frac{1}{\pi\omega} \right] \tag{11.24}$$

You may also be able to see that the transform of the impulse is

$$F[\delta(t)] = 1 \tag{11.25}$$

11.3.3 THE RECTANGLE FUNCTION

To demonstrate how to apply the Fourier Transform to a slightly more complex function, consider the rectangle function in Figure 11.5.

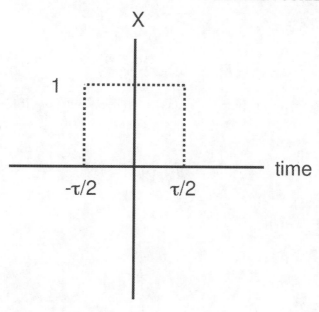

Figure 11.5: A rectangular function that has a value of 1 between $-\tau/2$ and $\tau/2$ and 0 everywhere else.

As you might suspect, it is going to take an infinitely many sinusoids to represent this signal because of the sharp corners and because it is not periodic. To start we can plug in the rectangle function into Equation 11.19.

$$F[Rect(t)] = \int_{-\tau/2}^{\tau/2} 1e^{-j\omega t}\,dt \tag{11.26}$$

Here we have changed the limits of integration because the rectangle function doesn't contribute anywhere else other than between $-\tau/2$ and $\tau/2$.

$$F[Rect(t)] = -\frac{1}{j\omega}e^{-j\omega t}\Big|_{-\tau/2}^{\tau/2} \tag{11.27}$$

$$F[Rect(t)] = \frac{e^{j\omega(\tau/2)} - e^{-j\omega(\tau/2)}}{j\omega} \tag{11.28}$$

and using Euler's Identify from Appendix A

$$e^{j\omega t} - e^{-j\omega t} = 2j\,\sin(\omega t) \tag{11.29}$$

so,

$$F[Rect(t)] = \tau \frac{\sin(\omega\tau)}{\omega\tau} \tag{11.30}$$

There is a special name for the function $\sin(x)/x$—the *sinc* function. It is shown graphically in Figure 11.6. So,

$$F[Rect(t)] = \tau\, sinc(\omega\tau) \tag{11.31}$$

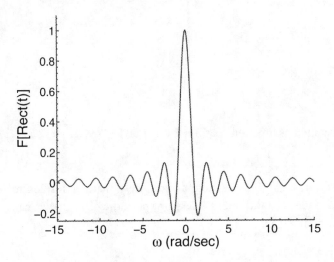

Figure 11.6: A sinc function as the Fourier Transform of the rectangle function in Figure 11.5.

In Figure 11.6, we can think of the values telling us the amplitudes of the various sinusoids (the constants in Equation 11.2) needed to represent the rectangle function. And, as expected, infinitely many sinusoids are needed to get a better and better representation of the sharp corners (i.e., the oscillations never die out). Also note that some of the terms are negative for some frequencies. Another caveat is that we have plotted both positive and negative ω, but in reality we would only need to represent the positive frequencies because the plot is symmetric.

11.3.4 INVERSE FOURIER TRANSFORM

Similar to the Laplace Transform we can define an inverse that will undo the Fourier Transform.

$$y(t) = \int_{-\infty}^{\infty} F(\omega)e^{j\omega t}\, d\omega \tag{11.32}$$

11.4   THE DISCRETE FOURIER TRANSFORM

In reality, biological signals do not have a compact analytical representation. In other words, we will not be able to write down a simple equation for $f(t)$. To further complicate matters, most biological signals are acquired by a computer system and are therefore *discrete* signals. A very efficient method of taking the Fourier Transform of a discrete and aperiodic signal is the Fast Fourier Transform (FFT). The algorithm is considered one of the most significant in the history of computing, because it is very simple and is much faster than computing the discrete Fourier Transform by brute force. We will not go into the details of how it works, but a bit of background is needed to understand how to use it and its limitations.

The basic formula for a Discrete Fourier Transform (DFT) is

$$X[n] = \sum_{m=1}^{M} x[m]e^{-j2\pi mn/M} \tag{11.33}$$

We will parse this a bit, but there are a few things we need to get sorted out first. $x[m]$ is our data set in time and there are samples from 1 to M. But where did we get this from? It is usually from some measurement device that *sampled* the data as it was coming in. The reason it was sampled is because we can't store an infinite number of values on a computer. What this means is that instead of taking every value at all possible times, we only take a value every so often. If we take data every 2 msec then we would say that our *sampling frequency* is 1/0.002 or 500Hz. One way to think about the value of 2msec is that it is the time distance between the samples m and $m + 1$.

We can also talk about the fundamental harmonic in Equation 11.33 as being

$$f_0 = \frac{f_s}{M} \tag{11.34}$$

because it is the frequency of the sinusoid that would span the entire data set. For example, if we collected 2 seconds of data at 500Hz we would have M=1000 datapoints.

$$f_0 = \frac{500Hz}{1000} = 0.5Hz \tag{11.35}$$

So the fundamental frequency would be half of a Hertz. And then the harmonics of that data set are just multiples of the harmonic.

$$f_n = n\frac{f_s}{M} \tag{11.36}$$

So we can now see that when we sample, we will not get perfect frequency resolution (as we do in the continuous Fourier Transform), but rather get some spacing between frequencies. And taking more points (sampling for a longer time period) will also impact the frequency resolution. It is important to recognize that in Equation 11.33, m is the index of the sample data and n controls the harmonic of the fundamental frequency.

The fundamental also tells us the frequency resolution

$$\Delta f = \frac{f_s}{M} \tag{11.37}$$

So in our example the frequency resolution will be 0.5Hz.

In Equation 11.33 the harmonics of the fundamental frequency are represented by the term $2\pi n/M$ (in rad/sec). You can now map the terms from the continuous Equation 11.19 to the discrete Equation 11.33.

11.4.1 ALIASING AND THE NYQUIST RATE

We run into an interesting problem with Equation 11.33. At the root of the problem is a fundamental equality

$$e^{j\omega_0 n} = e^{j(\omega_0 + 2\pi)n} \tag{11.38}$$

It may be easier to see why this is the case in the following

$$\cos\left[(\omega_0 + 2\pi)n\right] = \cos\left[\omega_0 n + 2\pi n\right] = \cos\left[\omega_0 n\right] \tag{11.39}$$

In other words, discrete frequencies separated by some integer multiple of 2π cannot be distinguished from one another. The somewhat non-obvious result is that when we compute Fourier Coefficients from the discrete Fourier Transform, they will correspond to real frequencies that have been *mapped* to the range of $-\pi$ to π (a span of 2π). So in a plot for the FFT we have squeezed all of the possible frequencies into some small range.

But there is still another catch. The frequencies are symmetric so the only unique values are from 0 to π. The value of π corresponds to the highest frequency we can represent. The key here is that the sampling rate and number of samples will set the range of real frequencies that can be represented.

Graphically we can understand why there is a limit to the highest frequency we can resolve by sampling the sinusoid in Figure 11.7 at two different rates. In the figure, you can see that if you sample at the top rate, you can resolve every peak and valley. But if you sampled at a lower rate, you cannot not resolve the frequency. Instead, you would report some much lower (and not correct) frequency.

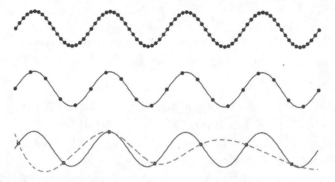

Figure 11.7: An example of sampling the same sinusoid at different rates. Note that in the bottom plot, where there is a low sampling rate, the result (green dashed line) would be the appearance of a lower frequency signal.

11.4.2 THE NYQUIST RATE AND ALIASING

The Nyquist Theorem states that the highest frequency that is possible to resolve is half of the sampling rate. That means that when we sampled as 500Hz above, we could then only resolve frequencies between 0 and 250Hz. That is because 0 to 250Hz will map to half of the frequency domain range from 0 to π.

You will often hear this stated in reverse. If you know that the highest frequency in a signal is, say, 300Hz, then you need to sample at a rate of at least 600Hz. We say *at least*, because this is the theoretical minimum sampling frequency. Graphically, this would be like sampling at the rate needed to put a point at each peak and valley of the sinusoid. Again, that would be ideal. But consider what would happen if you, just by chance, sampled at the same rate but at a different phase—you might end up hitting all of the zero crossings! So, in reality, you should sample much faster than the Nyquist Rate. A good rule of thumb would be to at least double the Nyquist rate.

So what happens to real frequencies that fall outside the range that can be resolved? They are simply some 2π multiple of a frequency that *is* inside the range. In other words, they contribute to some other (lower) frequencies that are in the range. And it would be impossible to distinguish how much of the contribution came from the lower frequency and how much came from the higher. This problem of higher frequencies contributing to the Fourier Coefficients of lower frequencies is known as *aliasing*. You can in fact see how that happens graphically in the bottom panel of Figure 11.7. The slow sampling rate means that high frequencies might contribute to low frequency coefficients.

11.5   MATLAB

As you might expect, Matlab has a number of functions that are designed to handle frequency data. We will only cover the most important and most used here—taking the FFT. Let us first create a sinusoid.

```
>> Fs = 1000; % Sampling Frequency in Hertz (Cycles/Second)
>> Ts = 1/Fs;  % Sampling Period (Seconds)
>> Time = 0:Ts:10;  % Create 10 seconds of data
>> f = [100 204 306];
>> Data = 1.5*cos(2*pi*f(1)*Time); % Data created in 3 steps
>> Data = Data + 1.75*sin(2*pi*f(2)*Time);
>> Data = Data + 2.0*sin(2*pi*f(3)*Time); % Done only to fit on one line
>> M = length(Data);  % Number of Points
```

From this code we could easily get the number of seconds present in some other ways, for example:

```
>> SecondsPresent = M*Ts;
```

Just to check, we can take a look at our time data

```
>> plot(Time,Data);
```

You should see data (you may need to zoom a bit) that is the sum of three sinusoids. Next we need to make our frequency vector. This will be the frequencies that we know we can resolve.

```
>> F = 0:Fs/M:Fs/2;
```

Note that our frequency resolution is just $\frac{F_s}{M}$ and the highest frequency we know we can resolve (the Nyquist frequency) is $F_s/2$. Now we can use the *fft* command in Matlab to get the full Discrete Fourier Transform.

```
>> FDataComplex = fft(Data);
```

The problem is that the transform returns complex values for the coefficients (i.e., real and imaginary numbers). What we want is the magnitude (the absolute value in Matlab).

```
>> FData = abs(FDataComplex);
```

Then we are ready to plot the data, but remember that we only want half of the data.

```
>> HalfFData = FData(1:length(F));
>> plot(F,HalfFData);
```

You should see that we have now fixed our original problem, as presented in Figure 11.2. Here, simply using the cross-correlation caused the frequencies 204Hz and 306Hz to be unresolved. But now we can see the peak due to these frequencies. You can use the code above as a template for taking the Fast Fourier Transform of any arbitrary signal just by changing the Data and Sampling Rate.

A last example will use the code above to demonstrate the idea of aliasing, using your sound card. You can download the Matlab function from the website called *aliasing.m*. You should read the help for the file and follow the suggested example.

11.6 EXERCISES

1. Give two examples of everyday items that need to report out a frequency of one type or another.

2. Find the two-sided Fourier transform (analytical derivation of $\int_{-\infty}^{\infty}$) of the signal

$$x(t) = e^{-a|t|}$$

You should assume that $a > 0$. Hint: You may need to treat $t < 0$ differently than $t > 0$. Show your work.

3. You all have a "second" brain called the enteric brain. It is a small network of neurons associated with your stomach that operates almost entirely independently from your central nervous system. While studying this fascinating collection of neurons you have collected an electrical signal for 30 seconds at a sampling rate of 1kHz. To analyze your signal you have calculated the Discrete Fourier Transform (DFT).

   a. What range of frequencies (bandwidth) can you represent? Explain.

   b. What frequency resolution will you have? Explain.

4. Download the file *eeg_data.mat* from the website. The file is a single channel recording of an EEG (downloaded from www.physionet.org), measured in units of μV. These data were sampled at 50Hz. You can load these data into Matlab using the *load* command.

   a. How many seconds of data are present?

   b. Use Matlab to plot the signal. Include labels. Axes should be in seconds and mV.

   c. What is the frequency resolution?

   d. What is the highest frequency you can trust?

   e. In Matlab, take the Fast Fourier Transform (see example code in the chapter) and plot the frequencies on appropriate axes with labels.

    f. Complex signals are often composed of many sinusoids. They may, however, contain bands of frequencies that contribute more to the signal than other band. Here, a band means some range of frequencies. For the EEG signal given, estimate the three dominant frequency ranges.

CHAPTER 12

Filters

If you have spent the night in a large city or tried to study while your dormmates were having a party, you have encountered the idea of noise. In both cases, what you would really like to do is get rid of the "noise." For now, let's simply define noise as anything that you don't want and a filter to be a way of getting rid of the noise. Of course if you are out for a night on the town, or just finished a final, the "noise" in the previous examples may be just what you are looking for. So in reality noise is in the eye of the beholder. Filters, in fact, are the basis for some cool products like noise-cancellation headphones. In this chapter we will explore two general ways in which filters are used to eliminate noise from a signal: 1) frequency filtering and 2) windowing.

A simple example comes from the theater. Stage lights typically only come in white, but to get different colors, stagehands will place a piece of cellophane over the light. The action of the cellophane is to cut out all of the light except for a specific color.

So far our examples have been for filters of sound waves or for light. But most physical systems can have filters. For example, the shocks on your car filter out slight fluctuations in the road. The result is a nice smooth ride. Filters are also very important in the medical and research world and in thinking about various sub-systems of your physiology. Of course your eyes and ears act on light and sounds, but you can think of your knees as simple shock absorbers. In your cells there are a number of very simple ion buffers that help soak up excess ions, or release them if there is not enough—a sort of chemical shock absorber. And a number of medical research techniques, like flow cytometry, are really just elaborate filtering processes.

So how does a filter actually work? In Chapter 10 we defined a signal as

$$x(t) = a(t) + b(t) + c(t) + \ldots + z(t) \tag{12.1}$$

where $q(t)$ was the signal we wanted and everything else was noise. In general, there is no way to separate out $q(t)$ without knowing more information about either q or the noise. In a very typical situation the information that we know would be about the frequencies present. For example, let's suppose that $q(t)$ is an EEG signal that has power between 2Hz and 110Hz (i.e., the power spectrum has values in this range). The most simple filter would then keep any signal in the range of 2-110Hz but eliminate all frequencies lower than 2Hz and higher than 110Hz. This might not get rid of all of the components in Equation 12.1, but it is a start. If any of our noise is outside the 2-110Hz range we will have effectively eliminated it! We will discuss what to do with noise that is in our frequency range of interest later.

The typical way of filtering out noise is to use a combination of *amplifying* the parts of the signal you want, $q(t)$, and *attenuating* the parts you don't want. To start quantifying how we can gain the right combination, we will think about filters as transfer functions.

12.1 IDEAL FILTERS

In Chapter 11, we developed the Bode plot as a way of showing how the amplification or attenuation of a system changes as the input frequency was changed. As this is exactly what we want to show for a filter, we will use Bode plots to characterize the transfer functions of our filters. Below in Figure 12.1 are the Bode plots of four generic types of *ideal* filters.

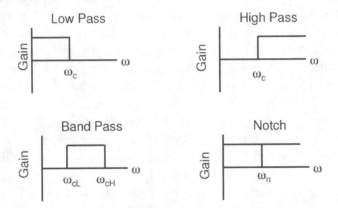

Figure 12.1: The bode plots for four types of ideal filters.

Low Pass Filters will allow through (dB(gain) = 0) all low frequencies and eliminate (dB(gain)<0) frequencies higher than some value. The *cut-off frequency* defines where the border is between the frequencies we want to keep and those we want to eliminate. Often the cut-off frequency is given the symbol, ω_c. Note that for an ideal lowpass filter we are keeping frequencies lower than ω_c (to the left on the Bode plot) and completely removing all higher frequencies. A good application of a lowpass filter is when we know that a biological system could not possibly generate a frequency over some range. For example, it is rare for a biological system to generate frequencies above 100Hz.

High Pass Filters will allow through (dB(gain) = 0) all high frequencies and eliminate (dB(gain)<0) frequencies lower than some value. Again we can use the idea of a cut-off frequency. This time we are keeping frequencies higher than ω_c (to the right on the Bode plot). A good biological application of a highpass filter is to remove what is known as *drift*. This is a slow change in the baseline of a signal, and is something that happens often in long-term monitoring of a signal. For example when using a gel electrode for electrical recordings from muscles (EMG), the contact between the skin and the electrode can change at different phases of respiration, leading to a slow drift in the baseline. This slow change can be taken out of the signal using a highpass filter.

Bandpass Filters allow through frequencies (dB(gain) = 0) over some range but eliminate (dB(gain)<0) frequencies outside that range. Here we must specify both the low (ω_c^L) and high (ω_c^H) cut-offs. All frequencies between the cut-offs are said to be in the *passband*. In our EEG example above we would want to have a bandpass filter with $\omega_c^l = 2Hz$ and $\omega_c^h = 110Hz$. But don't forget that Bode plots usually show frequency in *rad/sec*! Another way to think about a bandpass filter is as a combination of high and lowpass filters. In our example, the lowpass filter would allow through everything lower than 110Hz. Our highpass filter would allow through all frequencies higher than 2Hz. Two additional terms are sometimes used to quantify a bandpass filter. The *center frequency*, ω_0, is the frequency at the very center of the passband. The *bandwidth* is defined as the difference between the low and high cut-off frequencies. In our EEG filter example, $\omega_0 = 56Hz$ and the bandwidth would be 54Hz.

Notch Filters allow through *all* frequencies *except* over some range. Again, we need to define both low (ω_c^l) and high (ω_c^h) cut-offs. But here rather than allow through frequencies between the cut-offs, we eliminate them. Frequencies outside this range will be passed. In effect a notch filter is the opposite of a bandpass filter and is sometimes called a *bandstop*. In our EEG example, consider what happens if we had noise at 60Hz (a common source of noise coming from power outlets and lights). This falls right in the range 2-110Hz and our bandpass filter will pass it through with the rest of the signal. But we can add a notch filter with $\omega_c^l = 59Hz$ and $\omega_c^h = 61Hz$, to eliminate the noise. It is true that any EEG signal from 59-61Hz will also be eliminated, but that may be a small and acceptable price to pay for reduced noise.

12.1.1 IDEAL FILTER PHASE SHIFT

All Bode plots contain gain and phase information. If you think of a phase change as a delay, you will quickly see that an ideal filter would not have any delay (phase shift) at all. In this case, all of the phase plots would be a straight line at 0 degrees over the entire frequency range.

12.1.2 THE CHIRP SIGNAL

A good demonstration of the different types of filters is to send a *chirp* signal into the filter and observe the output. Briefly, the chirp signal is a sinewave that has a constantly increasing frequency over time (see Figure 12.2).

The signal is called a chirp because when played through speakers, it sounds like a bird chirp. Figure 12.3 shows the ideal output for each of the filters. Note that for ideal filters the chirp signal will abruptly begin (full amplitude=1) in the passband or end (amplitude=0) in the stopband.

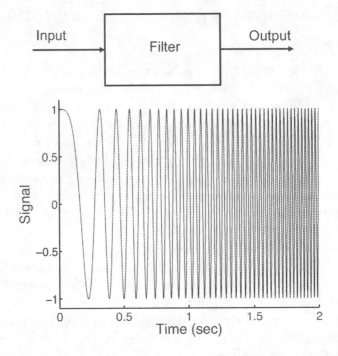

Figure 12.2: A filter as a system characterized by a transfer function. A *chirp* input that steadily increases in frequency from 0 to 30Hz.

12.2   FILTERS IN REALITY

Unfortunately, it is not possible to create a transfer function with perfect pass and stopbands. The reason is because with a finite number of poles we can never make the attenuation drop from 1 to 0 right at the cut-off frequency. There is always some less abrupt transition and we will always have some phase changes too.

Although we can never make a perfect filter, there are ways to engineer filters to fit desired specifications. It is worth noting, however, that many natural systems have transfer functions that are very effective filters. For example, the transfer functions for your ear and eye are very effective bandpass filters. Consider that both senses detect wavelengths (of sound or light) within some range. And the transfer function for a bat's hearing will be different than yours.

12.2.1   ROLL-OFF

Since we cannot achieve an ideal filter we would like to have some measures of the quality of a real filter. Figure 12.4 shows the gain portion of a generic Bode plot for a lowpass filter. We will discuss each element of this plot to build up some terminology. Although a lowpass filter is shown, the same terminology will be used for the other three types of filters.

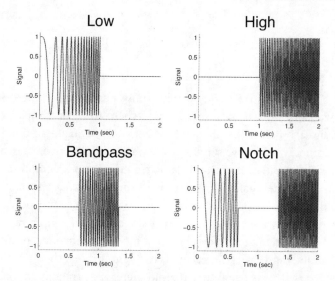

Figure 12.3: The ideal time response of the four filters in Figure 12.1 to the chirp input in Figure 12.2. The four panels correspond to the four filters. The lowpass filter allows through all frequencies from 0-15Hz, while the highpass allows only 15-30Hz. The Bandpass allows the frequencies between 10-20Hz, while the notch eliminates 10-20Hz.

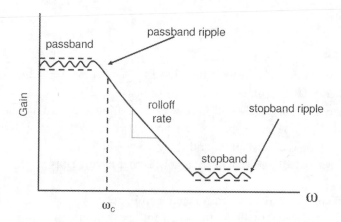

Figure 12.4: A typical Bode plot for a real filter.

Notice in Figure 12.4 that the gain plot no longer has a sharp transition at the cut-off frequency. In fact, it is difficult to know what we should call the cut-off. By convention the point where the signal power has dropped by $\frac{1}{2}$ is ω_c. If we compute this gain in dB we arrive at:

$$20 \log(\frac{1}{2}) = -3\text{dB} \tag{12.2}$$

So, as in Chapter 8, our cut-off frequency is at half power and is called the -3dB *point*. Therefore, on a Bode plot we can find ω_c as the frequency where the gain drops to -3dB. Remember, however, that the signal power is the square of the actual gain. So at half power our gain is actually $\sqrt{0.5} = 0.707$. In other words, if we send in a sinewave with a frequency of ω_c and an amplitude of 1, the output will be a sinewave with the same frequency but an amplitude of 0.707.

One way to assess the quality of a filter is to find the *roll-off* from the 3dB point. There are two measures that are used often (and they are related). In general, a fast fall-off (high slope) of the gain is good. Since we want to quantify a slope, we can simply take the number of dBs that are dropped over some range of frequencies. Since the frequency axis is typically plotted as $\log(\omega)$, the most natural (and easiest) measure is to see how much the gain drops for every integer multiple on the log axis. This is equal to a 10-fold increase in frequency. The name for a 10-fold increase is a *decade*, so it is common to characterize the rate of roll-off in dB/*decade*.

A second measure is dB/octave. An *octave*, like a decade, is just a way of comparing frequencies on the frequency axis. It is defined, the same as in music theory, as a doubling of frequency. A bit of thought will show that dB/octave and dB/decade are related (they are both just a slope measurement). In fact $3.\overline{3}$ dB/octave = dB/decade because it takes $3.\overline{3}$ doublings to get from 1 to 10.

12.2.2 RIPPLES

In the passband and stopbands some filters have *ripples*. These are variations in the gain as a function of frequency. The result is that even in the passband, some frequencies will have a gain of a bit more than 1 while others will have a gain of a bit less than 1. The passband ripple is typically reported as a value in dB that reflects how much above (or below) 0 the gain is.

Ideally there would be no ripples and the gain would be 1 for the entire passband. So why include ripples? The tradeoff, as we will see below, is that having ripples allows for a faster roll-off rate.

In the stopband we can never completely eliminate the signal (i.e., the gain never actually goes to 0). Instead, what we usually see is more and more attenuation of the signal to the point where it is effectively removed.

12.2.3 PHASE SHIFTS

There is no generic phase terminology for a filter. But that doesn't mean that the phase isn't an important consideration when designing or evaluating a filter. Let's first consider passing two sinewaves into an ideal filter:

$$x(t) = \sin(\omega_1 t) + \sin(\omega_2 t) \tag{12.3}$$

if ω_1 and ω_2 are in the passband then the ideal output is the same as the input:

$$y(t) = \sin(\omega_1 t) + \sin(\omega_2 t) \tag{12.4}$$

Now let's consider passing the same two sinewaves into a filter where the gain is still 1 for ω_1 and ω_2 but the phase shifts are 30 and 160 degrees respectively. The output *should* be

$$y(t) = \sin(\omega_1 t + 30\pi/180) + \sin(\omega_2 t + 160\pi/180) \tag{12.5}$$

But in a real filter it won't be! The reason is that different frequency components are not lining up nicely in the Time Domain. Another way to think about this is that the filter will delay the information in any given frequency component by some amount. And the information at one frequency may be delayed differently than the information in another frequency.

For the two sinewaves above we might be able to untangle the phase shifts. But consider what happens if the gains were different at ω_1 and ω_2. Furthermore, consider what would happen if we input a real biological signal composed of many frequencies—they will all be shifted by different amounts.

There is no simple way to characterize filters based on phase shifts but some general heuristics are helpful to consider. Since we are going to attenuate our signal in the stopband anyway, the phase shifts for those frequencies don't matter much; we are most concerned with any phase alterations that might be made in the passband. So the best option would then be to have no phase shift in our passband. The next best option is if the phase shift is the same (e.g., always 60 degrees) in the entire passband. Here we at least know that the relative phases of the different frequencies in our signal are being preserved. Again, this is not always possible. The next best option is if the phase varies linearly over the passband. The worst case is when the phase shift is some non-linear function or has discontinuous jumps.

12.3 FIRST AND SECOND ORDER FILTERS

Another way to characterize a filter is by the *filter-order*, which is simply the number of poles of the transfer function. Below we will examine first and second-order filters and discuss some of their properties.

12.3.1 A FIRST ORDER FILTER

The most basic filter is an RC electric circuit or mechanical spring and dash-pot system. In general, any system that has a single element that can store a quantity will act as some sort of first-order filter. In Chapter 8 we considered the transfer function for a simple high and lowpass RC filter ($R = 10k\Omega, C = 2\mu F$).

$$H_L(s) = \frac{1}{RCs + 1} \tag{12.6}$$

$$H_H(s) = \frac{RCs}{RCs + 1} \tag{12.7}$$

And to these equations we substituted $s = j\omega$ to get at the frequency response. You can now see that all we have done here is move from the s-domain to the frequency domain.

$$H_L(j\omega) = \frac{1}{RCj\omega + 1} \tag{12.8}$$

$$H_H(j\omega) = \frac{RCj\omega}{RCj\omega + 1} \tag{12.9}$$

Where $1/RC$ is the cut-off frequency ($50 rad/sec$). Figure 12.5 shows the frequency response of both low and highpass filters (corresponding to Figures 8.11 and 8.13). And we see that they do fall off the way we expected ($-3dB$ at the cut-off). We will also see that rolloff rate (described below) is 6dB/decade.

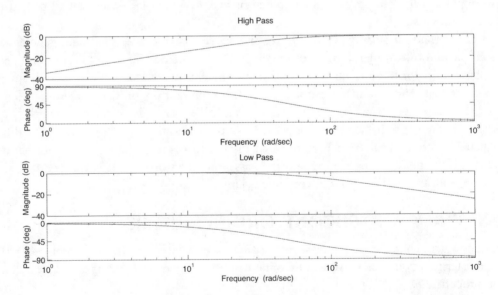

Figure 12.5: The Bode plot for the high and lowpass RC filters.

Since this is our first encounter with a real filter, we can send in a chirp signal to see what happens. The result is shown in Figure 12.6. Remember back to the lowpass filter in Figure 8.12, where we put in various frequency sinewaves. You will remember that as we turned up the frequency of the input, the output became smaller and smaller. And this is exactly what the RC circuit does to the chirp signal. But from Figure 12.5 we can see that, unlike in Figure 12.3, there is no longer a sharp roll-off. Instead there is a gradual transition between the stop and pass bands. So we start out seeing something that looks a bit like a sinewave, but the high frequencies are quickly smeared out and give something that levels out to a straight line at 0.

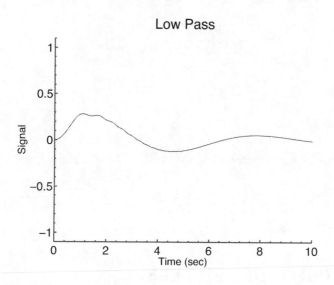

Figure 12.6: The filtering of a chirp signal by a high and lowpass RC filter.

12.3.2 A SECOND ORDER FILTER

We can now turn to the second-order equation we considered in Chapter 8

$$H(s) = \frac{1}{s^2 + 0.5s + 1} \tag{12.10}$$

where we determined that this filter is a lowpass. But we also noticed that there were some frequencies that were amplified (e.g., gain greater than 0dB). And we defined the peak of this effect (at 1rad/sec) to be at the *resonant frequency*.

We can also define what is known as the *Q factor* to quantify the width of the frequencies over which the resonance occurs. The typical definition is to measure the gain in the passband and the gain at the peak. We then find the two points on either side of the peak that cross half way between the passband and peak gains. Then the Q factor is defined as the range of frequencies, $\Delta\omega$. In general, the Q factor measures how sensitive the system is to resonating at a particular

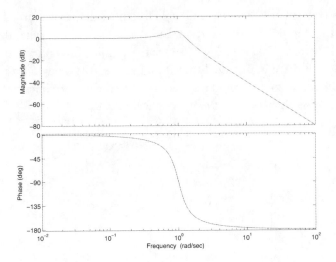

Figure 12.7: Bode plot for the second-order system.

frequency. If Q is small (small $\Delta\omega$) then the range is very small and the system only resonates for some small range of frequencies.

12.4   HIGHER ORDER FILTERS

Since a filter is nothing more than a transfer function, we can design our transfer function (placement of poles and zeros) to achieve some desired effect. A general rule of thumb is that increasing the number of poles will increase the fall-off. For each pole you will generally add 6dB/decade of roll-off. So for a transfer function with one pole (first-order filter) the fall-off will be 6dB/decade. For an 8 pole (8th-order) filter the fall-off will be 48dB/decade. The down side of adding many poles is that they introduce negative phase. Adding zeros on the other hand will have little effect on the fall-off but will introduce positive phase. Therefore a good filter will somehow balance zeros and poles to produce a small phase shift.

Although you could develop your own method of placing poles and zeros, there are a number of generic filters that have already been designed. The difference in these filters is largely the location of the poles and zeros and therefore the gain and phase response. Below we will show some examples of higher-order filters for lowpass filters.

12.4.1   BUTTERWORTH

The *Butterworth* filter is defined by a transfer function with poles in the left half plane (stability) that fall along a circle in the complex plane (see Figure 12.8). Mathematically

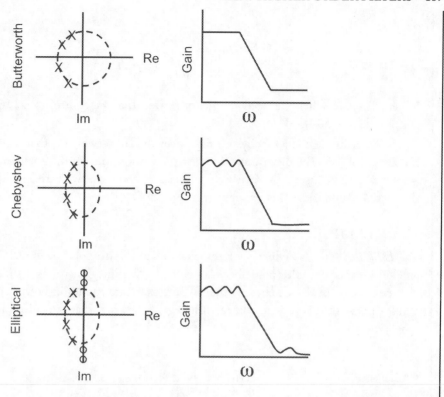

Figure 12.8: Generic filters, showing the location of poles and zeros of the transfer functions on the left and the Bode plot on the right.

$$G(\omega) = \frac{1}{\sqrt{(1 + \omega/\omega_c)^{2n}}} \qquad (12.11)$$

where $2n$ is the filter order and ω_c is the cut-off frequency (-3dB point). The reason it is $2n$ is because we want to add pairs of complex poles. Increasing n will cause the fall-off to increase and will flatten the passband. The trade-off is that it will sacrifice other important filter parameters (such as phase shifts), to get a flat passband.

12.4.2 CHEBYSHEV

The *Chebyshev* filter is defined by a transfer function with poles in the left half plane that form an ellipse in the complex plane.

$$G(\omega) = \frac{1}{\sqrt{1 + \epsilon^2 C_n^2\left(\frac{\omega}{\omega_c}\right)}} \tag{12.12}$$

Where $C_n()$ is a Chebyshev polynomial of degree n, and ϵ controls the amplitude of the ripples (amplitude = $\frac{1}{\sqrt{1+\epsilon^2}}$).

The advantage of a Chebyshev filter over the Butterworth is that for the same order, there is a faster roll-off. The disadvantage is that there are ripples in the passband and the phase shift is somewhat non-linear. The Chebyshev filter also comes in two varieties. Type I is for ripples in the passband and Type II is for ripples in the stopband.

12.4.3 ELLIPTICAL

The *Elliptical* filter is defined by a transfer function with poles in the left half plane that form an ellipse in the complex plane but also zeros on the imaginary axis. In fact, the elliptical filter is a special case of the Chebyshev filter where there are ripples in both the pass and stop bands and the phase shift is very non-linear. The big advantage is that the fall-off is even faster than the Chebyshev filters.

$$G(\omega) = \frac{1}{\sqrt{(1 + \epsilon^2 R_n^2\left(\xi, \frac{\omega}{\omega_c}\right)}} \tag{12.13}$$

Where $R_n()$ is the Chebyshev rational function of the nth-order, ϵ controls the passband ripple and a combination of ϵ and ξ (selectivity factor) control the stopband ripple. As the ripple in the stopband goes to zero, the Elliptical filter becomes a Chebyshev filter.

12.4.4 BESSEL

The *Bessel* filter (also called a Thomson filter) is defined the same way as the Chebyshev filter but where $C_n()$ is replaced by a Bessel polynomial, $B_n()$. The advantage of the Bessel filter is that the phase shift is flat in the passband and linear in the transition and stopbands. The result is very good time-domain properties. The disadvantage is that for the same order the roll-off is not as steep as the previous three filter types.

12.4.5 FILTER EVALUATION

We can now take a closer look at how the various filters compare to one another. First, we will examine how changing the order of a filter will make the roll-off occur faster. Figure 12.9 shows a first, second, third and fourth-order lowpass Butterworth filter with a cut-off frequency of $250 rad/sec$. With each increase in order (addition of a pole to the transfer function) we gain an extra 6db of roll off per decade. Something to note about the Butterworth filter is that a first-order Butterworth filter is actually just an RC circuit!

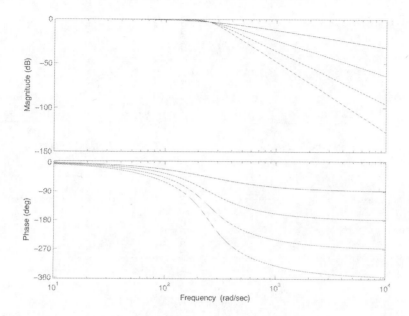

Figure 12.9: Bode plots of Butterworth filters of first, second, third and fourth-order.

In Figure 12.10 is a comparison of the four different types of filters for the same order and cut-off frequency. The point here is that each has certain strengths and weaknesses that can be taken advantage of (or avoided) in certain situations.

12.4.6 HIGH, BANDPASS AND NOTCH FILTER

All of the examples above were for lowpass filters. There is, however, a simple algorithm to convert any lowpass filter to any other type of filter. First, a lowpass filter, F(s), is created where $\omega_c = 1 rad/sec$. Then F(s) is replaced with the expressions in the table below.

Type	Equation	Notes
Highpass	$F(\frac{\omega_c}{s})$	ω_c is the desired cut-off
Bandpass	$F(\frac{s^2+\omega_0^2}{sBW})$	ω_0 is center frequency, BW is desired bandwidth
Notch	$F(\frac{sBW}{s^2+\omega_0^2})$	ω_0 is center frequency, BW is desired bandwidth

If we make the substitutions into the lowpass Butterworth filter we can generate the transfer functions, and Bode plots (shown in Figure 12.11) for the four different types of filters.

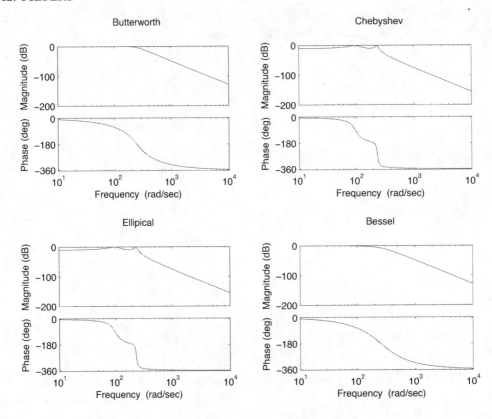

Figure 12.10: The four different types of higher-order filters (4th-order), all as lowpass with a cut-off of $250 rad/sec$.

12.4.7 ELECTRICAL IMPLEMENTATION

It may seem that adding more poles (higher-order) would always be good design practice. However, keep in mind that each pole is the same as adding an additional storage element. So in a real filter this will mean more parts (i.e., capacitors). And with the extra components will come consideration for cost, heat dissipation, size, weight or other considerations.

The most popular implementation of a filter is the second-order (2-pole) *active* filter (contains an operational amplifier) know as a *Sallen-Key* filter. It is very simple (see Figure 12.12) and has a number of very nice properties. Perhaps the most useful is that we can string together a number of 2-pole filters in series to create higher-order filters.

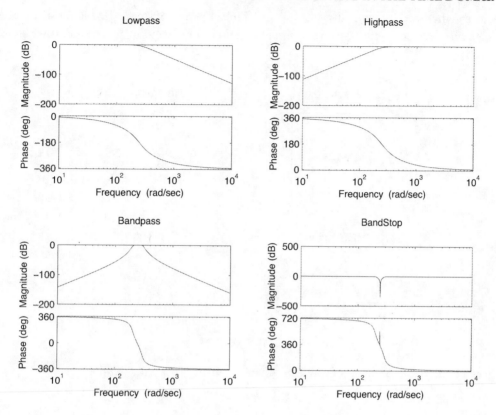

Figure 12.11: Fourth-order Butterworth filters of various types.

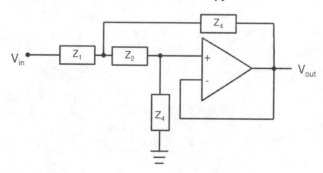

Figure 12.12: A generic Sallen-Key filter where the Z_{1-4} may be resistors or capacitors depending upon the type of filter (e.g., highpass, lowpass) desired.

12.5   WINDOWING IN THE TIME DOMAIN

We took a quick look at what filters can do in the Time Domain using the chirp signal, but for the most part, the design of our filters has occurred in the frequency domain. There are some instances where we implement a filter in the Time Domain instead. This is most often called *windowing*. The idea is that we have a signal and then some window function that we slide over the signal.

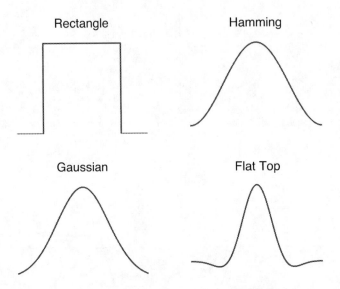

Figure 12.13: Some examples of windows. There are many other types out there.

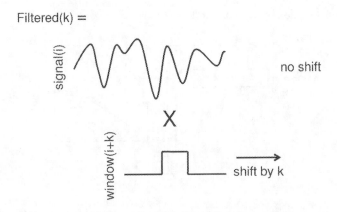

Figure 12.14: An example of using the cross-correlation with a window to smooth a signal.

Figure 12.13 shows some types of windows, but there are many others that are used, each with their own particular uses in signal processing. Windows are applied in much the same way as computing the cross-correlation with no time delay.

$$Filtered(k) = \frac{1}{N} \sum_{i=1}^{N} Signal(i)Window(i+k) \tag{12.14}$$

The result of applying the rectangle window to the signal in Figure 12.14 will be to *smooth* out the signal. So what the rectangle function does is to replace each point with the average of a number of nearby points. The number of points that will be averaged depends on the width of the window. A bit of thought will reveal that any big jumps will get averaged out. Now if we think about this in the frequency domain, what we have really done is to get rid of any high frequencies (the source of the jumping around), and so many windows are lowpass filters. The difference in the type of filter is just how the averaging is done. In the rectangle window all points within the window are given the same weight. But, in something like the Gaussian window points close by are given more weight. And in the Flat Top window we even give some points a negative weight. This is to make the center points even more pronounced.

We can even extend the one dimensional windowing idea to two dimensions to create image filters. In this case, they can be used to smooth out features, but also to bring out edges and features that are hard to see in the raw image. In both one and two dimensional windowing, you will sometimes hear the window referred to as a *kernel*. Something to note is that each of these Time Domain filters has some analog in the frequency domain. But they don't map perfectly. In other words, it wouldn't make sense to try to design a time-domain window that corresponds to a Butterworth filter.

12.6   MATLAB

Matlab has an entire toolbox dedicated to filters. It also has a generic commands called *filter* that will create a digital filter. In this section, however, we will cover some built in commands that will create the filters discussed in this chapter. The commands *butter*, *cheby1* (and *cheby2*) and *ellip* all take in certain parameters as inputs (e.g., the order, pass and stop band cut-off frequencies, ripple information, and type of filter). The output is the coefficients for the numerator and denominator of the transfer function. So a basic series of commands for creating a filter are

```
>> Cutoff = 250; %Cutoff at 250 rad/sec
>> [AHigh,BHigh]=butter(4,Cutoff,'high','s');
>> GHigh = tf(AHigh,BHigh);
>> bode(GHigh);
```

Here we have created a fourth-order, highpass Butterworth filter with a cut-off frequency of 250 rad/sec. Then to apply the filter, we can simply use the *lsim* command to send an input signal into the filter transfer function.

```
>> t = 0:0.0001:1;
```

```
>> f = [50 250 500];
>> V = 1.5*cos(2*pi*f(1)*t); %only split to keep in margins
>> V = V+ 1.75*sin(2*pi*f(2)*t);
>> V = V+ 2.0*sin(2*pi*f(3)*t);
>> [Out,t]=lsim(GHigh,V,t);
>> plot(t,Out,'k'); %Signal after highpass
```

More options and examples can be found in the file *FilterExamples.m*, posted online.

12.7 EXERCISES

1. Find examples in the student cafeteria where filters are being used.

2. As explained in the chapter, a chirp signal is very useful to an engineer looking to characterize real system (or filter) since it will sweep through a range of frequencies. You will create a particular chirp that will increase from 0-400Hz over a period of 10 seconds. You will then apply a number of filters and note the effect.

   a. Plot a chirp signal over time. Save the commands below into Matlab script

   ```
   Fs = 10000; %Sampling Frequency of 10kHz
   Time = 0:1/Fs:10; %Generate a time vector of 10 seconds
   %Chirp signal of frequencies ranging from 0 to 400Hz
   ChirpSignal = chirp(Time,0,10,400);
   ```

   b. Listen to the chirp signal on your computer using the following command

   ```
   >> sound(ChirpSignal,Fs);
   ```

   c. Create a lowpass Butterworth filter of order 4 with a cut-off of 100Hz (remember to convert to rad/sec before creating your filter). Apply this filter to the chirp signal using the *lsim* command in Matlab. You must turn in your Matlab commands as well as a plot of the filter output and explain the results. You should listen to the signal using the *sound* command.

   d. Create a highpass Butterworth filter of order 4 with a cut-off of 200Hz. Apply this filter to the chirp signal. You must turn in your Matlab commands as well as a plot of the filter output and explain the results. You should listen to the signal using the *sound* command.

   e. Create a bandpass Butterworth filter of order 4 with a low cut-off of 100Hz and a high cut-off of 200Hz. Apply this filter to the chirp signal. You must turn in your Matlab commands as well as a plot of the filter output and explain the results. You should listen to the signal using the *sound* command.

f. Create a bandstop Butterworth filter of order 4 with a low cut-off of 100Hz and a high cut-off of 200Hz. Apply this filter to the chirp signal. You must turn in your Matlab commands as well as a plot of the filter output and explain the results. You should listen to the signal using the *sound* command.

3. Matlab has some built-in datasets that can be used for demonstration purposes. One dataset is a clip of Handel's Hallelujah Chorus. You can load these data into Matlab by typing

```
>> load handle
```

Two variables will appear in your workspace, Fs (sampling frequency) and y (audio data). You can listen to the clip by typing:

```
>> sound(y,Fs)
```

a. Plot the audio signal on an appropriate time axis (take into account the sampling frequency).

b. Take the FFT of this signal, again taking into account the sampling frequency. Show both your Matlab commands and your plot. Taking an FFT of a signal with unknown frequency content is a good first step toward designing an appropriate filter. You should notice that there are two dominate frequency peaks that roughly separate the lower instruments and voices from the higher instruments and voices.

c. Design a Chebyshev type 1 filter (order 4) to only pass the low instruments and voices. Show your Matlab commands and explain your choice of filter type (low, high, band or stop) and cut-off frequency. You will also need to supply the filter with a value for the peak-to-peak ripple in the passband. You should use a value of 1dB.

d. Apply your lowpass filter and show a plot of the new audio file. Also show a plot the FFT of your new signal to verify that you have filtered the high frequencies.

e. Listen to your new clip to be sure you have filtered out the high frequencies. You may notice that there is a slight echo that was introduced. Explain where this echo came from.

4. You have recorded your first EEG in lab (download data *NoisyEEGData.mat* from the website, Sampling Frequency of 200Hz) and are about to show your instructor your results. As you are walking down the hall you find out from an EE that for the past few days they have been practicing sending 10Hz and 13Hz signals through antennas as part of their class. You know that your EEG content is in the range from 0-25Hz. A quick check of your data shows that your signals have been contaminated by their signals. This is a problem because you expect to be taking many more EEGs throughout the semester and will need to find a solution to this problem.

a. Plot the original signal on appropriate time axes.

b. Take the FFT of your noisy signal to observe the peaks at 10 and 13Hz. Turn in this plot.

c. Your first instinct is to rerecord your data but with a simple lowpass RC circuit filter added. This will eliminate some of the high frequency you want but at least you will get the lower frequencies. Create a software version of this filter to find out if this will be helpful. Explain your choice of R, C and cut-off frequency. Plot your signal after being passed through this filter as well as the FFT. Did it work?

d. Design an elliptic filter that will only stop frequencies in the range being sent by the EEs. View the results of this filter. Show your Matlab commands and explain your choices for cut-off frequencies, ripple and order. Plot the signal after being passed through the elliptic filter you have created as well at the FFT. Did it work?

e. Design a Butterworth filter that will only stop frequencies in the range being sent by the EEs. View the results of this filter. Show your Matlab commands and explain your choices for cut-off frequencies and order. Plot the signal after being passed through the Butterworth filter you have created as well at the FFT. Did it work?

f. Compare the filters you have created. Although not necessary, it may be helpful to compare the Bode plots of your filters. Which do you think is best and why? Which would cost the least to implement in hardware?

APPENDIX A

Complex Numbers

A.1 INTRODUCTION

Complex numbers do not mean to be difficult. In fact, they are not too bad once you get used to their somewhat peculiar behavior in mathematical equations. In reality they are only complex because they are made up of two parts, one real and one imaginary.

You have heard all about real numbers (0, 1, -3.4, π and $\sqrt{2}$). But the square root of -1 ($\sqrt{-1}$) has been given the name *imaginary*. Although mathematicians and physicists use the letter i to denote the imaginary number, engineers typically use j to avoid any potential confusion with current (i, or I).

There is only one imaginary number, j, because we can express any imaginary value as the product of the imaginary number and a real number. For example, we can represent $\sqrt{-64} = \sqrt{64}\sqrt{-1} = 8j$. Although it is not the traditional interpretation, you can think of j as a kind unit and the real value as the magnitude.

A complex number, C, is simply a number that has both a real part and an imaginary part.

$$C = A + Bj \tag{A.1}$$

where A is a real number and Bj is an imaginary number. The important part is that both A and B are the normal real numbers we are used to, but B has been multiplied by the imaginary number j. Some examples are 5 + 2j, or 2π + 3j.

A.2 THE COMPLEX PLANE

An alternative way to express a complex number is to use the terms *phase* and *magnitude*. To do so we can introduce a graphical representation of a complex number in the *complex plane*. The complex plane is simply an xy axis where the real part of a complex number is on the x-axis and the imaginary part of a complex number is on the y-axis.

In this way, we can plot a point on the complex plane to represent a complex number. For example, to graphically represent the point 5 +2j, we would place a point on the complex plane at (5,2).

Once a complex number is represented in the complex plane, we can draw a line (vector) from the origin (0,0), to the point. The length of this line is the magnitude and is defined as:

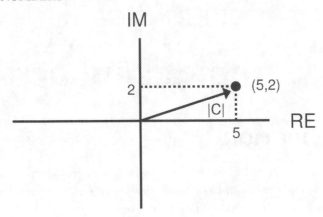

Figure A.1: The complex plane along with the graphical representation of 5+2j and the magnitude $|C|$.

$$|C| = \sqrt{(real\ part)^2 + (imaginary\ part)^2} = \sqrt{5^2 + 2^2} = \sqrt{104} \qquad (A.2)$$

The angle our vector makes with the x-axis is the *phase*. and can be computed as

$$< C = \tan^{-1}\left(\frac{imaginary\ part}{real\ part}\right) = \tan^{-1}(2/5) = 0.3805\text{rad} \qquad (A.3)$$

Typically the phase is represented in *radians*, often abbreviated as *rad*. The conversion between radians and degrees is

$$radians = \frac{\pi(\text{degrees})}{180} \qquad (A.4)$$

so the above value can be converted

$$\text{degrees} = 0.3805\frac{180}{\pi} = 21.801\text{degrees} \qquad (A.5)$$

A.3   EULER'S IDENTITY

Euler's Identity is one of the most interesting relationships in all of mathematics since it shows a powerful relationship between five important numbers.

$$e^{j\pi} + 1 = 0 \qquad (A.6)$$

A bit of substitution and trig identities reveals another interesting identity

$$e^{j\pi} = \cos\theta + jsin\theta \tag{A.7}$$

or alternatively

$$\cos(\theta) = \frac{e^{j\theta} + e^{-j\theta}}{2} \tag{A.8}$$

$$\sin(\theta) = \frac{e^{j\theta} - e^{-j\theta}}{2j} \tag{A.9}$$

Going back to our vector diagram we can now find a *polar form* for a complex number.

$$C = A + Bj = r(\cos\theta + jsin\theta) = re^{j\theta} \tag{A.10}$$

where $re^{j\theta}$ is a complex number in polar form. Note that r is the magnitude and θ is the phase. We can also define the following relationships.

$$A = Re[C] = rcos\theta \tag{A.11}$$

$$B = ImC = rsin\theta \tag{A.12}$$

It may seem like a waste of time to express a complex number in polar form. But consider that the solution to linear differential equations takes the form of exponentials raised to the poles (eigenvalues). Therefore, when the poles are complex numbers, the polar form makes life much easier.

A.4   MATHEMATICAL OPERATIONS

It was mentioned above that performing mathematical operations on complex numbers is not hard, but that it requires a bit of attention. Below we show a few operations that are valuable in signals and systems.

A.4.1   ADDITION AND SUBTRACTION

Adding or subtracting complex numbers is performed by adding or subtracting the real and imaginary parts separately. For example

$$(2 + 3j) - (4 + j) = -2 + 2j \tag{A.13}$$

A.4.2   MULTIPLICATION

Multiplying complex numbers is the same as multiplying binomial expressions. Keep in mind that $j^2 = -1$. For example

$$(2 + 3j)(4 + j) = 8 + 2j + 12j + 3j^2 = 5 + 14j. \qquad \text{(A.14)}$$

A.4.3   CONJUGATION

In linear systems complex poles always occur in pairs, $A \pm Bj$. These two solutions are *complex conjugates*. For example, if $C = A + Bj$, the $C^* = A - Bj$, where $*$ denotes the complex conjugate. It is interesting that multiplying a pair of complex conjugates by one another always results in a real number.

APPENDIX B

Partial Fraction Expansion

In deriving transfer functions in the Laplace (s) domain, we often arrive at an equation of the form

$$G(s) = \frac{(s - z_1)(s - z_2)...(s - z_m)}{(s - p_1)(s - p_2)...(s - p_n)} \tag{B.1}$$

and can sometimes be difficult to interpret. Instead it would be helpful to deal with a number of more simple transfer functions

$$G(s) = \frac{a_1}{(s - p_1)} + \frac{a_2}{(s - p_2)} + ... + \frac{a_n}{(s - p_n)} \tag{B.2}$$

Here the interpretation is that we are transforming a single block into many smaller blocks in series:

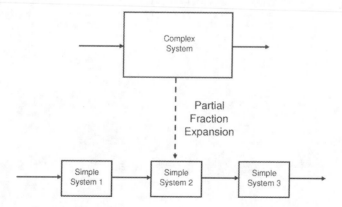

Figure B.1: Graphical representation of a partial fraction expansion.

The method to perform this is called a *partial fraction expansion*. We can find the poles by simply factoring the denominator. So what we need to find are the constants $a_{1,...,n}$. The mechanical way to do this is

$$a_i = (s - p_i)G(s)|_{s=p_i} \tag{B.3}$$

We repeatedly use this equation to find all of the a constants. This is somewhat abstract so let's look at an example. Given the following transfer function

$$G(s) = \frac{8s^2 + 3s - 21}{s^3 - 7s - 6} \qquad (\text{B.4})$$

we must first find the poles. Doing so yields poles at -2, 3 and -1. Next we would like to rewrite our transfer function as:

$$G(s) = \frac{a_1}{s + 2} + \frac{a_2}{s - 3} + \frac{a_3}{s + 1} \qquad (\text{B.5})$$

To find a_1 we use equation B.3:

$$a_1 = (s + 2)G(s)|_{s=-2} \qquad (\text{B.6})$$

or

$$a_1 = \frac{8s^2 + 3s - 21}{(s - 3)(s + 1)}|_{s=-2} = 1 \qquad (\text{B.7})$$

Likewise

$$a_1 = (s - 3)G(s)|_{s=3} \qquad (\text{B.8})$$

or

$$a_2 = \frac{8s^2 + 3s - 21}{(s + 2)(s + 1)}|_{s=3} = 3 \qquad (\text{B.9})$$

and

$$a_3 = (s + 1)G(s)|_{s=-1} \qquad (\text{B.10})$$

or

$$a_3 = \frac{8s^2 + 3s - 21}{(s - 3)(s + 2)}|_{s=-1} = 4 \qquad (\text{B.11})$$

so our partial fraction expansion leads to

$$G(s) = \frac{1}{s+2} + \frac{3}{s-3} + \frac{4}{s+1}$$

(B.12)

There is a more complex process that must be followed for repeated poles which will not be covered here.

APPENDIX C

Laplace Transform Table

$x(t)$ (time)	$x(s)$ (laplace)	Notes
$\delta(t)$	1	Impulse
$u(t)$	$\frac{1}{s}$	Step
$af(t)$	$aF(s)$	Scaling
$f_1(t) + f_2(t)$	$F_1(s) + F_2(s)$	Additivity
$f(t - t_0)$	$F(s)e^{-st_0}$	Time Shift
$e^{-at}f(t)$	$F(s+a)$	S-domain Shift
$\frac{d^n f}{dt^2}$	$s^n F(s) - f^n(0) - f^{n-1}(0)$ $\ldots - f^0(0)$	Derivative
$\int_0^\infty f(t)dt$	$\frac{1}{s}F(s)$	Integral
$f(at)$	$\frac{1}{|a|}F(\frac{s}{a})$	Time Expansion/Contraction
t	$\frac{1}{s^2}$	Ramp
e^{-at}	$\frac{1}{s+a}$	Exponential Decay/Growth
te^{-at}	$\frac{1}{(s+a)^2}$	Time Varying Decay/Growth
$\frac{1}{a}(1 - e^{-at})$	$\frac{1}{s(s+a)}$	
$e^{-a} - e^{bt}$	$\frac{b-a}{(s+a)(s+b)}$	
$\frac{1}{ab} - \frac{e^{-at}}{a(b-a)} - \frac{e^{-bt}}{b(a-b)}$	$\frac{1}{s(s+a)(s+b)}$	
$\frac{1}{a^2}(1 - e^{-at} - ate^{-at})$	$\frac{1}{s(s+a)^2}$	
$(1 - at)e^{-at}$	$\frac{s}{s(s+a)^2}$	
$\sin(bt)$	$\frac{b}{s^2-b^2}$	Pure Sine
$\cos(bt)$	$\frac{s}{s^2-b^2}$	Pure Cosine
$e^{-at}\sin(bt)$	$\frac{b}{(s+a)^2+b^2}$	Decaying/Growing Sine
$e^{-at}\cos(bt)$	$\frac{s+a}{(s+a)^2+b^2}$	Decaying/Growing Cosine

APPENDIX D

Fourier Transform Table

$f(t)$ (time)	$F()$ (frequency)	Notes
$f_1(t) + f_2(t)$	$F_1(\omega) + F_2(\omega)$	Addition
$k\,f(t)$	$kF(\omega)$	Multiplication
$f(at)$	$\frac{1}{\|a\|}F(\frac{\omega}{a})$	Time Scaling
$F(t)$	$2\pi f(-\omega)$	Transform Symmetry
$f(t - t_0)$	$F(\omega)e^{-j\omega t_0}$	Time Shift
$f(t)e^{-j\omega t_0}$	$F(\omega - \omega_0$	Frequency Shift
$f_1(t) * f_2(t)$	$F_1(\omega)F_2(\omega)$	Time Convolution
$f_1(t)f_2(t)$	$\frac{1}{2\pi}F_1(\omega) * F_2(\omega)$	Frequency Convolution
$\frac{d^n f}{dt^n}$	$(j\omega)^n F(\omega)$	Time Differentiation
$\int_t^{-\infty} f(t)dt$	$\frac{F(\omega)}{j\omega} + \pi F(0)$	Time Integration
$u(t)$	$\pi\delta(t) + \frac{1}{j\omega}$	Step Function
$\mathrm{rect}(\frac{t}{\tau})$	$\tau\,\mathrm{sinc}(\frac{\tau\omega}{2})$	Rectangle Function
$\frac{W}{2\pi}\mathrm{sinc}(Wt)$	$\mathrm{rect}(\frac{\omega}{2W})$	Sinc Function

Author's Biography

JOSEPH TRANQUILLO

Joseph Tranquillo is an associate professor of biomedical engineering at Bucknell University where he has been a faculty member since 2005. He received his Doctor of Philosophy degree in biomedical engineering from Duke University (Durham, NC) and Bachelor of Science degree in engineering from Trinity College (Hartford, CT). He was the founder and inaugural chair of the Biomedical Engineering Society Undergraduate Research Track, co-chair of the Body-Of-Knowledge task force, and is currently the program chair of the American Society for Engineering Education (ASEE) Biomedical Engineering Division (BED). He co-founded the KEEN Winter Interdisciplinary Design Experience and is the co-director of the Bucknell Institute for Leadership in Technology and Management. Joe has received funding from the NSF, NIH, NCIIA, KEEN, and the Department of Defense and his work has been featured on the Discovery Channel, TEDx and CNN Health. He has published over 50 technical and 85 engineering education proceedings and articles and is the author of two other textbooks, *Quantitative Neuroelectrphysiology* and *Matlab for Engineering and the Life Sciences*. Joe has won the 2010 National ASEE BED teaching award, Bucknell's Presidential Teaching Award in 2013 and is a National Academy of Engineering Frontiers of Engineering Education faculty member. When not teaching or doing research, he enjoys improvisational dance and music, running trail marathons, backpacking, brewing Belgian beers, and raising his two children Laura and Paul.

Printed in the United States
by Baker & Taylor Publisher Services